STATIONARY ENGINEER

HIGH PRESSURE PLANT TENDER

HIGH PRESSURE
BOILER OPERATING ENGINEER

STATIONARY ENGINEER

HIGH PRESSURE PLANT TENDER

HIGH PRESSURE BOILER OPERATING ENGINEER

Harry Mahler, P.E.

PRENTICE
HALL
PRESS

New York   London   Toronto   Sydney   Tokyo   Singapore

Sixth Edition

 PRENTICE HALL PRESS

Simon & Schuster, Inc.
15 Columbus Circle
New York, NY 10023

An Arco Book
Published by Prentice Hall Press

Prentice Hall Press and colophons are
registered trademarks of Simon & Schuster, Inc.

Manufactured in the United States of America

3   4   5   6   7   8   9   10

Library of Congress Cataloging-in-Publication Data
Main entry under title:

Stationary engineer, high pressure boiler operating
engineer, high pressure plant tender.

  1. Steam engineering—Examinations, questions, etc.
2. Steam engineering—Problems, exercises, etc.
I. Mahler, Harry. II. Hammer, Hy.
TJ280.5.S73   1986   621.1'6'076   85-11139
ISBN 0-668-06075-1 (Paper Edition)

CONTENTS

WHAT THIS BOOK WILL DO FOR YOU

ARCO Publishing, Inc. has followed testing trends and methods ever since the firm was founded in 1937. We specialize in books that prepare people for tests. Based on this experience, we have prepared the best possible book to help *you* score high.

To write this book we carefully analyzed every detail surrounding the forthcoming examination . . .

- the job itself
- official and unofficial announcements concerning the examination
- all the previous examinations, many not available to the public
- related examinations
- technical literature that explains and forecasts the examination

CAN YOU PREPARE YOURSELF FOR YOUR TEST?

You want to pass this test. That's why you bought this book. Used correctly, your "self-tutor" will show you what to expect and will give you a speedy brush-up on the subjects tested in your exam. Some of these are subjects not taught in schools at all. Even if your study time is very limited, you should:

- Become familiar with the type of examination you will have.
- Improve your general examination-taking skill.
- Improve your skill in analyzing and answering questions involving reasoning, judgment, comparison, and evaluation.
- Improve your speed and skill in reading and understanding what you read—an important part of your ability to learn and an important part of most tests.

This book will tell you exactly what to study by presenting in full every type of question you will get on the actual test.

This book will help you find your weaknesses. Once you know what subjects you're weak in you can get right to work and concentrate on those areas. This kind of selective study yields maximum test results.

This book will give you the *feel* of the exam. Almost all our sample and practice questions are taken from actual previous exams. On the day of the exam you'll see how closely this book follows the format of the real test.

This book will give you confidence *now*, while you are preparing for the test. It will build your self-confidence as you proceed and will prevent the kind of test anxiety that causes low test scores.

This book stresses the multiple-choice type of question because that's the kind you'll have on your test. You must not be satisfied with merely knowing the correct answer for each question. You must find out why the other choices are incorrect. This will help you remember a lot you thought you had forgotten.

After testing yourself, you may find that you are weak in a particular area. You should concentrate on improving your skills by using the specific practice sections in this book that apply to you.

PART ONE

Applying and Studying For Your Job

STUDYING AND USING THIS BOOK

Even though this course of study has been carefully planned to help you get in shape by the day your test comes, you'll have to do a little planning on your own to be successful. You'll also need a few pointers proven effective for many other good students.

SURVEY AND SCHEDULE YOUR WORK

Regular mental workouts are as important as regular physical workouts in achieving maximum personal efficiency. They are absolutely essential in getting top test scores, so you'll want to plan a test-preparing schedule that fits in with your usual program. Use the Schedule on the next page. Make it out for yourself so that it really works with the actual time you have at your disposal.

There are five basic steps in scheduling this book for yourself and in studying each assignment that you schedule:

1. SCAN - the entire job at hand.
2. QUESTION - before reading.
3. READ - to find the answers to the questions you have formulated.
4. RECITE - to see how well you have learned the answers to your questions.
5. REVIEW - to check up on how well you have learned, to learn it again, and to fix it firmly in your mind.

SCAN

Make a survey of this whole book before scheduling. Do this by reading our introductory statements and the table of contents. Then leaf through the entire book, paying attention to how the Background Study Material relates to the Sample Examinations. When you have this bird's eye view of the whole, the parts take on added meaning, and you'll see how they will help improve your score.

QUESTION

As you scan, questions will come to your mind. Write them into the book. Later on you'll be finding the answers. For example, in scanning this book you would naturally change the headline *Studying and Using This Book* into "What don't I know about studying? What are my good study habits? How can I improve them? How should I go about reading and using this book?" Practice the habit of formulating and writing such questions into the text.

READ

Now, by reviewing your questions you should be able to work out your schedule easily. Stick to it. And apply these five steps to each assignment you give yourself in the schedule. Your reading of each assignment should be directed to finding answers to the questions you have formulated and will continue to formulate. You'll discover that reading with a purpose will make it easier to *remember* the answers to your questions.

RECITE

After you have read your assignment and found the answers to your questions, close the book and recite to yourself. For example, if your question here was "What are the five basic steps in attacking an assignment?" then your answer to yourself would be scan, question, read, recite, and review. Thus, you check up on yourself and "fix" the information in your mind. You have now seen it, read it, said it, and heard it. The more senses you use the more you learn.

REVIEW

Even if you recall your answers well review them in order to "overlearn." "Overlearning" gives you a big advantage by reducing the chances of forgetting. Definitely provide time in your schedule for review. It's the clincher in getting ahead of the crowd. You'll find that "overlearning" won't take much time with this book because the text portions have been written as concisely and briefly as possible. You may be tempted to stop work when you have once gone over the work before you. This is wrong because of the ease with which memory impressions are bound to fade. Decide for yourself what is important and plan to review and overlearn those portions. Overlearning rather than last minute cramming is the best way to study.

Plan to study difficult subjects when you can give them your greatest energy. Some people find that they can do their best work in the early morning hours. On the other hand, it has been found that forgetting is less when study is followed by sleep or recreation. Plan other study periods for those free times which might otherwise be wasted . . . for example lunch or when traveling to and from work.

Plan your schedule so that not more than 1½ or 2 hours are spent in the study of any subject at one sitting. Allow at least a half-hour for each session with your book. It takes a few minutes before you settle down to work.

You will find that there is enough time for your study and other activities if you follow a well-planned schedule. You will not only be able to find enough time for your other activities, but you will also accomplish more in the way of study and learning. A definite plan for study increases concentration. If you establish the habit of studying a subject at the same time each day, you will find that less effort is required in focusing your attention on it.

Where To Study

1. Select a room that will be available each day at the same time. This will help you concentrate.
2. Use a desk or table which will not be shared so that you can "leave things out." It should be big enough to accommodate all your equipment without cramping you. Eliminate ornaments and other distractions.
3. Select a room that has no distractions, and keep it that way.
4. Provide for good air circulation in your study room.
5. Keep the temperature around 68°.
6. Provide adequate lighting . . . use a desk lamp in addition to overhead lights.
7. Noise distracts, so keep radio and TV turned off.
8. Arrange to have a permanent kit of necessary study equipment . . . pen, pencil, ruler, shears, eraser, notebook, clips, dictionary, etc.

Study On Your Own

As a general rule you will find it more beneficial to study with this book in your room alone. There are times, however, when two or more individuals can profit from team study. For example, if you can't figure something out for yourself you might get help from a friend who is also studying for this test. Review situations sometimes lend themselves to team study if everyone concerned has already been over the ground by himself. Sometimes you can gain greater understanding of underlying principles as you volley ideas back and forth with other people. Watch out, though, that you don't come to lean on the others so much that you can't work things out for yourself.

PROVEN STUDY SUGGESTIONS

1. Do some work every day in preparation for the exam.
2. Budget your time—set aside a definite study period for each day during the week.

3. Study with a friend or a group occasionally—the exchange of ideas will help all of you. It's also more pleasant getting together.
4. Answer as many of the questions in this book as you can. Some of the questions that you will get on your actual test will be very much like some of the questions in this book.
5. Be physically fit. Eat the proper food—get enough sleep. You learn better and faster when you are in good health.
6. Take notes.
7. Be an active learner. Participate. Try harder.

TECHNIQUES OF EFFICIENT STUDY

1. Do not attempt serious study while in too relaxed a position.
2. Avoid serious study after a heavy meal.
3. Do something while studying . . . make notes, underline, formulate questions.
4. Begin concentrating as soon as you sit down to study. Don't fool around.
5. Make time for study by eliminating needless activities and other drains on your precious time.
6. Make up your own illustrations and examples to check on your understanding of a topic.
7. Find some practical application of your newly acquired knowledge.
8. Relate newly acquired knowledge to what you knew before.
9. Consciously try to learn, to concentrate, to pay attention.
10. Look up new words in your dictionary.

Concentrating

Most students who complain that they don't know how to concentrate deserve no sympathy. Concentration is merely habit and ought to be as readily acquired as any other habit. The way to begin to study is simply to begin.

Don't wait for inspiration or for the mood to strike you, nor should you permit yourself to indulge in thoughts like, "This chapter is too long" or "I guess I could really let that go until some other time."

Such an attitude throws an extra load on your mental machinery, and by making you work against a handicap, makes it harder for you to begin.

Reading aloud is a good device for those whose minds wander while studying. Articulating "subvocally" for a few moments is another tonic for drifting thoughts. If this doesn't work, write down the point you happen to be dealing with when your mind "goes off track."

Do your studying alone, and you'll find it much easier to concentrate. If you are certain you need help on doubtful or difficult points, check these points and list them; you can go back or ask about them later. In the meantime, proceed to the next point.

A "little tenseness" is a good thing because it helps you keep alert while studying. Do without smoking, or newspapers, or magazines, or novels which may lead you into temptation. Studying in one place all the time also helps.

Boiling it all down, the greatest asset for effective studying is plain, garden variety "common sense" and will power.

Grasshoppers Never Learn

Don't be a "skipper." Jumping around from one part of your course to another may be more interesting, but it won't help you as much as steady progress from the start to the finish.

Studying and learning takes more than just reading. The "text" part of your course can be a valuable tool in test-preparation if you use it correctly. Reread the paragraph that gives you trouble. Be certain that you understand it before you pass on to the next one. Many persons who have been away from school for a long time, and those people who have a habit of reading rapidly, find that it helps if they hold a piece of white paper under the paragraph they are reading, covering the rest of the page. That helps you concentrate on the facts you are absorbing. Keep a pen or pencil in your hand while reading, and underline important facts. Put a question mark after anything that isn't quite clear to you, so that you can get back to it. Summarize ideas in the margins of your book. You'll be surprised how much easier it is to remember something once you have written it down, and expressed it in your own words.

THE KIND OF WORK
YOU WILL BE DOING

STATIONARY ENGINEERS

Nature of the Work

Stationary engineers operate, maintain, and repair the machinery that provides mechanical and electrical power for industry. They are also responsible for the equipment that heats, air-conditions, refrigerates, and ventilates factories and other buildings. The equipment they tend and control includes boilers, diesel engines, turbines, generators, pumps, condensers, and compressors. Much of it is similar to the nonstationary equipment operated by locomotive and marine engineers.

Stationary engineers start up and shut down equipment in order to meet demands for power and to insure the equipment is operating within established limits. They monitor meters, gauges, and other instruments attached to equipment and make adjustments whenever necessary. They also keep a log of all relevant facts about the operation and maintenance of the equipment. On a steam boiler, for example, they observe, control, and keep records of steam pressure, temperature, water level, power output, and the amount of fuel consumed. Stationary engineers control the flow of fuel to the boiler and the steam pressure by adjusting throttles, valves, or automatic controls.

Stationary engineers must periodically remove from equipment the soot and corrosion that can reduce operating efficiency. They test boiler water and add necessary chemicals to prevent corrosion and build up of deposits in the boiler.

These workers detect, identify, and correct any trouble that develops. They watch and listen to their machinery and routinely check safety devices. Often stationary engineers use hand or power tools to make repairs, ranging from a complete overhaul to replacing defective valves, gaskets, or bearings.

In a large plant, the stationary engineer may be in charge of the powerplant or engine room and direct the work of assistant stationary engineers, turbine operators, boiler tenders, and air-conditioning and refrigeration operators and mechanics. In a small plant, the stationary engineer may be the only person operating and maintaining equipment.

Working Conditions

Stationary engineers generally have steady year-round employment. They usually work a 5-day, 40-hour week. In plants that operate around the clock, they may be assigned to any one of three shifts—often on a rotating basis—and to Sunday and holiday work.

Engine rooms, powerplants, or boiler rooms usually are clean and well lighted. Even under the most favorable conditions, however, some stationary engineers are exposed to high temperatures, dust, dirt, and high noise levels from the equipment. General maintenance duties may cause contact with oil and grease, and fumes or smoke. Workers are on their feet a lot; they also may have a crawl inside boilers and work in crouching or kneeling positions to inspect, clean, or repair equipment.

Because stationary engineers work around boilers and electrical and mechanical equipment, they must be alert to avoid burns, electric shock, and injury from moving machinery.

Employment

Stationary engineers held about 58,000 jobs in 1982. They work in a wide variety of places, including factories, hospitals, schools, office and apartment buildings, shopping malls, hotels, and power stations. Usually, plants that operate on three shifts employ four to eight stationary engineers, but some have more. In many plants, only one engineer works on each shift.

Because stationary engineers work in so many different kinds of industries, they are employed in all parts of the country. Although some are employed in small towns and in rural areas, most work in the more heavily populated areas where large industrial and commercial businesses are usually located.

Training, Other Qualifications, and Advancement

Many stationary engineers start as helpers or oilers and acquire their skills through informal on-the-job experience. They usually become boiler tenders before advancing to stationary engineers. A good background also can be obtained in the Navy or Merchant Marine. However, most training authorities recommend formal apprenticeship programs because of the increasing complexity of the machines and systems; programs are sponsored by the International Union of Operating Engineers and the International Brotherhood of Firemen and Oilers, the principal unions to which stationary engineers belong.

In selecting apprentices, most local labor-management apprenticeship committees prefer high school or trade school graduates who have received instruction in mathematics, mechanical drawing, machine-shop practice, physics, and chemistry. Mechanical aptitude, manual dexterity, and good physical condition also are important qualifications.

The apprenticeship usually lasts 4 years. In addition to on-the-job training, apprentices receive classroom instruction in practical chemistry, elementary physics, blueprint reading, applied electricity, instrumentation, electronics, and other technical subjects.

Becoming a stationary engineer without going through a formal apprenticeship program usually takes many years of experience as an assistant to a licensed stationary engineer or as a boiler tender. This practical experience can be supplemented by technical or other school training or home study.

Many states and cities have licensing requirements for stationary engineers. Although requirements differ from place to place, applicants usually must be at least 18 years of age, reside for a specified period in the state or locality in which the examination is given, meet the experience requirements for the class of license requested, and pass a written examination.

Generally, there are several classes of stationary engineer licenses. Each class specifies the steam pressure or horsepower of the equipment the engineer can operate without supervision. The first class license permits the stationary engineer to operate equipment of all types and capacities. An applicant for this license may be required to have a high school education and an approved apprenticeship or on-the-job training. The lower class licenses limit the capacity of the equipment the engineer may operate without the supervision of a higher rated engineer.

Because of regional differences in licensing requirements, a stationary engineer who moves from one state or city to another may have to pass an examination for a new license. However, the National Institute for Uniform Licensing of Power Engineers has helped 21 states adopt a standardized licensing program that eliminates this problem by establishing reciprocity of licenses.

Stationary engineers advance to more responsible jobs by being placed in charge of larger, more powerful, or more varied equipment. Generally, engineers advance to these jobs as they obtain higher class licenses. Advancement, however, is not automatic. For example, an engineer who has a first-class license may work for some time as a boiler tender or an assistant to another first-class engineer before a vacancy occurs. Some stationary engineers eventually advance to jobs as plant engineers and as building and plant superintendents. A few obtain jobs as examining engineers and technical instructors.

Job Outlook

Little change in employment of stationary engineers is expected through the mid-1990's despite expanding industrial needs for more mechanical and electrical power. Employment in the occupation remained fairly constant through the 1970's due to automation and the shift to more powerful and centralized equipment. These trends will continue to limit growth in the future. Nevertheless, many job openings will arise because of the need to replace experienced workers who transfer to other occupations, retire or leave the labor force for other reasons.

Earnings

Stationary engineers have average hourly earnings that are almost 40 percent higher than the average for all nonsupervisory workers in private industry, except farming.

The principal unions to which these workers belong are the International Union of Operating Engineers and the International Brotherhood of Firemen and Oilers.

Related Occupations

Other workers who monitor and operate stationary machinery include nuclear reactor operators, power station operators, wastewater treatment plant operators, waterworks pump-station operators, chemical operators, and refinery operators.

Sources of Additional Information

Information about training or work opportunities is available from local offices of state employment services, locals of the International Union of Operating Engineers, and from state and local licensing agencies.

Specific questions about the occupation may be referred to:

International Union of Operating Engineers
1125 17th St. N.W.
Washington, D.C. 20036

National Association of Power Engineers, Inc.
176 West Adams St.
Chicago, Ill. 60603

For questions concerning licensing requirements, contact:

National Institute for Uniform Licensing of
Power Engineers
1436 Fritz Rd.
Verona, Wis. 53593

HIGH PRESSURE PLANT TENDER (BOILER TENDER)

Nature of the Work

Boiler tenders operate and maintain the steam boilers that power industrial machinery and heat factories, offices, and other buildings. They also may operate waste-heat boilers that burn trash and other solid waste.

Boiler tenders control the mechanical and or automatic devices that regulate the flow of air and fuel into the combustion chambers. They may, for example, start the pulverizers or stokers to feed coal into the firebox or start the oil pumps and heaters to ignite burners or sequence the safe burning of natural gas.

These workers inspect and maintain boiler equipment. Their work includes monitoring me-

ters and gauges attached to the boilers to ensure safe operation. Sometimes boiler tenders make minor repairs, such as packing valves or replacing indicators.

Boiler tenders also chemically test and treat water for purity to prevent corrosion of the boiler and buildup of scale.

Boiler tenders generally work under the supervision of licensed stationary engineers.

Working Conditions

Modern boiler rooms usually are clean and well lighted. However, boiler tenders may be exposed to noise, heat, grease, fumes, and smoke, and may have to work in awkward positions. They also are subject to burns, falls, and injury from defective boilers or moving parts, such as pulverizers and stokers. Modern equipment and safety procedures, however, have reduced accidents.

Employment

Almost one-half of the 62,000 boiler tenders employed in 1980 worked in factories. Plants that manufacture lumber, iron and steel, paper, chemicals, and stone, clay and glass products are among the leading employers of boiler tenders. Many others work in hospitals, schools, and office and apartment buildings. A large number also work in government agencies.

Although boiler tenders are employed in all parts of the country, most work in the more heavily populated areas where large industrial and commercial establishments are located.

Training, Other Qualifications, and Advancement

Some large cities and a few states require boiler tenders to be licensed. An applicant can obtain the knowledge and experience to pass the license examination by first working as a helper in a boiler room. Applicants for helper jobs should be in good physical condition and have mechanical aptitude and manual dexterity. High school courses in mathematics, motor mechanics, chemistry, and blueprint reading also are helpful to persons interested in becoming boiler tenders.

There are two types of boiler tenders' licenses—for low pressure and high pressure boilers. Tenders with licenses for low pressure boilers operate boilers generally used for heating buildings. Tenders of high pressure boilers operate the more powerful boilers and auxiliary equipment used to power machinery in factories as well as heat large buildings, such as high-rise apartments. However, tenders may operate equipment of any pressure if a licensed stationary engineer is on duty.

Due to regional differences in licensing requirements, a boiler tender who moves from one state or city to another may have to pass an examination for a new license. However, the National Institute for Uniform Licensing of Power Engineers is currently assisting many state licensing agencies in adopting uniform licensing requirements that would establish reciprocity of licenses.

Boiler tenders may advance to jobs as stationary engineers. To help them advance, they sometimes supplement their on-the-job training by taking courses in chemistry, physics, blueprint reading, electricity, and air-conditioning and refrigeration. Boiler tenders also may become maintenance mechanics.

Job Outlook

Little change in employment of boiler tenders is expected through the 1980's as more new boilers are equipped with automatic controls. Nevertheless, many openings will result each year from the need to replace experienced tenders who transfer to other occupations or retire.

Earnings

Boiler tenders have average hourly earnings that are higher than the average for all nonsupervisory workers in private industry, except farming.

The principal unions to which boiler tenders belong are the International Brotherhood of Firemen and Oilers and the International Union of Operating Engineers.

Related Occupations

Boiler tenders monitor and check steam boiler equipment which generates power for industrial machinery. Others whose work requires a similar background and related duties are oilers, operating engineers, power engineers, and stationary engineers.

Sources of Additional Information

Information about training or work opportunities in this trade is available from local offices of state employment services, locals of the International Brotherhood of Firemen and Oilers, locals of the International Union of Operating Engineers, and from state and local licensing agencies.

Specific questions about the nature of the occupation, training, and employment opportunities may be referred to:

National Association of Power Engineers, Inc.
176 West Adams St.
Chicago, Ill. 60603

International Union of Operating Engineers
1125 17th St. N.W.
Washington, D.C. 20036

EXAMINATION ANNOUNCEMENTS

STATIONARY ENGINEER

(Note: Although this announcement is for Stationary Engineers in New York City, examination announcements all across the country are very similar.)

Requirements

License and Certificate Requirements: By the last date of receipt of applications, candidates must possess a valid License for High Pressure Boiler Operating Engineer issued by the New York City Department of Buildings.

For appointment in certain New York City agencies, candidates may also be required to possess a valid Certificate of Equivalent Instruction issued by the New York City Department of Environmental Protection and/or a valid Certificate of Qualification for Refrigerating Machine Operator (Unlimited Capacity) issued by the New York City Fire Department. This license and certificates must be presented at the time of investigation and to the appointing officer at the appointment interview. In addition to the aforementioned license and certificates, candidates may also be required, within 90 days from the day of appointment, to obtain a valid Standpipe Certificate and Automatic Sprinkler Certificate issued by the New York City Fire Department.

Job Description

Duties and Responsibilities: Under supervision, operates, maintains and adjusts steam power plant and electro-mechanical building equipment; performs related work.

Examples of Typical Tasks: Operates, maintains and adjusts coal or oil-fired boilers, refuse burning furnaces, engines, pumps, heat exchangers, generators, motors, air conditioners, ejectors, gas scrubbers, fans, heating, ventilating, lighting and like equipment installed in public buildings, municipal pumping stations or incinerator buildings. Makes periodic inspections, minor repairs to equipment. Performs preventive maintenance. Takes responsible charge of a watch and, while so engaged, is responsible for and directs subordinate personnel. Prepares requisitions for materials and spare parts. When assigned, instructs and trains subordinates and other job-related personnel. Reads and logs meters, gauges and other recording devices. Keeps records and makes reports.

Test Information

Tests: The examination consists of an education and experience rating with a weight of 100. A grade of 70% is required. The education and experience ratings will be based on candidate's statements on Experience Paper Form A. There will be no other competitive test. Therefore, Experience Paper Form A must be filled out completely and in detail, including dates and number of hours worked per week.

Candidates are advised to list all education and experience for which credit may be given as described below since they will not be permitted to add additional education or experience once they have filed the application. Education and experience will not be found acceptable unless it is verifiable. Education and experience listed on Experience Paper Form A will receive credit only to the extent that it is described clearly and in detail.

You will receive a score of 70% for meeting the minimum requirements. After the minimum requirements are met, credit will be given on the following basis:

70-79.9   Experience as a stationary engineer or marine engineer (steam). Up to 1 point per year, depending on quality, will be given for each year of acceptable full-time paid experience at this level, up to a maximum of 79.9.

80-89.9   Experience as a chief engineer, or chief marine engineer (steam) supervising stationary engineers or marine engineers (steam). A score of 80 will be given for 1 year of full-time paid experience at this level. Up to 2 points per year, depending on quality, will be given for each additional year of acceptable full-time paid experience at this level, up to a maximum of 89.9.

90-100   Experience in charge of either the engineering operation of a large group of buildings including a power plant, or the operation of a public utility power house. After meeting the requirements for 80 above, a score of 90 will be given for 1 year of full-time paid experience at this level. Up to 2 points per year, de-

pending on quality, will be given for each additional year of full-time paid experience at this level, up to a maximum of 100.

In addition to the above:
(a)   4 points will be granted for possession of a baccalaureate degree in Mechanical Engineering or closely related engineering field issued upon completion of a course of study approved by the Accreditation Board for Engineering and Technology.
(b)   3 points will be granted for possession of a valid New York City Certificate of Qualification for Refrigerating Machine Operator (Unlimited Capacity).

The maximum rating that can be granted on this education and experience test is 100%.

NOTE: Marine Engineer (Steam) experience will be acceptable only if obtained while possessing a valid license issued by the United States Coast Guard Marine Inspection Service.

HIGH PRESSURE PLANT TENDER

(Note: Although this announcement is for High Pressure Plant Tenders in New York City, examination announcements all across the country are very similar.)

matic Sprinkler Certificate and Interior Fire Alarm Certificate issued by the New York City Fire Department.

Promotion Opportunities

Employees in the title of High Pressure Plant Tender who possess a valid License for High Pressure Boiler Operating Engineer issued by the New York City Department of Buildings and a valid Certificate of Equivalent Instruction issued by the New York City Department of Air Resources are accorded promotional opportunities, when eligible, to the title of Stationary Engineer. However, in addition to possessing the above license and Certificate, some City agencies may also require a valid Certificate for Refrigerating Machine Operator (Unlimited Capacity), a valid Standpipe Certificate, Auto-

Requirements

Minimum Requirements:
(1)   2 years of full-time paid experience within the last 10 years operating and maintaining high pressure boilers or the equivalent marine experience; or
(2)   not less than 1 year of such experience plus sufficient training of a relevant nature acquired in an approved trade or vocational high school to make up a total of 2 years of creditable experience. 6 months of acceptable experience will be credited for each year of approved trade or vocational high school.

The minimum requirements must be met by the last date for the receipt of applications.

Applicants may be summoned for the qualifying written test prior to the determination of whether they meet the minimum requirements.

Job Description

Duties and Responsibilities: Under direct supervision, tends coal or oil-fired high-pressure boilers and incinerator furnaces and related equipment, and assists in the maintenance and repair of this equipment; performs related work.

Examples of Typical Tasks: Hand fires high pressure boilers using solid fuels, and stoker equipment and incinerator furnaces using burnable refuse. Cleans fires. Tends and operates stoker equipment, oil-fired high-pressure boilers and incinerator furnaces and related equipment. Assists in the cleaning, inspection, testing, maintenance, and repair of high-pressure boilers, incinerators, auxiliaries and related equipment.

Test Information

Tests: A qualifying written multiple-choice test; and a practical-oral test, weight 100. A score of at least 70% is needed in each test in order to pass.

The qualifying written multiple-choice test may include questions on the operation and maintenance of high-pressure boilers, auxiliaries, and related equipment; operation and maintenance of oil and coal firing equipment; combustion controls, meters and gauges; use of appropriate tools, packing and gaskets; related mathematics, reading comprehension and safety; and other related areas. Your score on this qualifying written test will not affect your final score or your rank on the eligible list.

The practical-oral test may include questions on boiler operation and appurtenances; boiler maintenance and safety; firing and combustion; auxiliaries; meters and gauges and other related areas.

LICENSE ANNOUNCEMENTS

STATIONARY ENGINEER (ELECTRIC)

Job Description

Duties and Responsibilities: Under general supervision, operates, inspects, and adjusts high and/or low voltage electrically powered plant equipment including diesel engines; performs related work.

Examples of Typical Tasks: Operates, inspects and adjusts diesel engines, generators, motors, transformers, converters, rectifiers, controllers, switchboards, circuit breakers, pumps, compressors, sewage treatment process units, etc. Oils, cleans, makes minor repairs and supervises the maintenance work of others on this equipment. Supervises the taking and analysis of process samples as required by the National Pollution Discharge Elimination System permits. Reads meters, gauges and recording devices. Keeps records. Makes reports. May be placed in responsible charge of a watch.

Requirements

(1) 5 years of full-time paid experience within the last 10 years in responsible charge of the operation of high tension electrical plants; or

(2) 2 years of such experience and a bachelor's degree in electrical or mechanical engineering from an accredited college; or

(3) 2 years of such experience plus three years of full-time paid experience as a journeyman electrician; or

(4) A satisfactory equivalent.

Test Information

A written test will be of the multiple-choice type and may include questions on fundamentals of electrical circuits and machines; operation and maintenance of electric motors, tion and maintenance of electric motors, generators, and auxiliary equipment; operation and maintenance of mechanical equipment such as internal combustion engines, pumps, compressors, and auxiliary equipment; supervision and record keeping, safety; and related areas.

Applicants who pass the written test will be required to pass a qualifying practical oral test. The practical-oral test may include questions on electric services, switching and protective equipment; electrical control panels; electrical machinery; maintenance, testing and repairs; and electrical diagrams.

Applicants will be required to pass qualifying medical and physical tests prior to appointment.

Tests

Written, weight 100, 70% required; practical-oral, qualifying, 70% required. The written test will be of the multiple-choice type and may include questions on fundamentals of electrical circuits and machines; operation and maintenance of electric motors, generators, and auxiliary equipment; operation and maintenance of mechanical equipment such as internal combustion engines, pumps, compressors, and auxiliary equipment; supervision and record keeping, safety and related areas. Candidates who pass the written test will be required to pass a qualifying practical-oral test conducted at an electrically-powered plant. The qualifying practical-oral test may include questions on electric services, switching and protective equipment; electrical control panels; electrical machinery; maintenance, testing and repairs; and electrical diagrams.

Promotion Opportunities

Employees in the title of Stationary Engineer (Electric) are accorded promotion opportunities, when eligible, to the title of Senior Stationary Engineer (Electric).

HIGH PRESSURE PLANT TENDER

(Note: Although this License Announcement enables the applicant to practice in New York City, License Announcements all across the country are very similar.)

This is NOT an application for a City position. It is an application for a High Pressure Boiler Operating Engineer License. Possession of this license qualifies the holder to operate boilers that carry a pressure of more than 15 pounds of steam per square inch and are rated in excess of 10 hp. or if such boilers produce hot water at a pressure of 160 psi or at a temperature over 250°F.

Print or type the information requested fully and accurately, giving exact names, dates and addresses. Detection of falsity in any material statement will result in your disqualification.

LICENSE FOR HIGH-PRESSURE BOILER OPERATING ENGINEER
(Local Law No. 76 of 1968 as Amended)

Applications: Issued and received continuously during the normal hours of business of the Application Section.

The General Regulations for License Examinations of the New York City Department of Personnel apply to all license or permit examinations given by the Department of Personnel. The General Regulations for License Examinations are available in the Application Section of the Department of Personnel, 49 Thomas Street, Manhattan. Applicants shall consult the General Regulations for License Examinations and the specific notice of examination for the license or permit for which they are applying, and shall be responsible for knowledge of their contents.

Requirements: Applicants must meet the following qualifications at the time of filing for this examination:

1. Be at least 18 years of age;
2. Be able to read and write the English language; and
3. Have been employed as a fireman, oiler, general assistant, journeyman boiler maker or a machinist to a licensed high pressure boiler operating engineer in a building or buildings in the City of New York for a period of 5 years of the 7 years immediately preceding the date of his application; or
4. Have received the degree of mechanical engineer from a school or college recognized by the University of the State of New York, and have had 1 year's experience in the operation and maintenance of high pressure boilers under the supervision of a licensed high-pressure boiler operating engineer in the City of New York within the 7 years immediately preceding the date of his application; or
5. Have been a holder for a period of at least 4 years of a certificate as engineer issued by a board of examining engineers duly established and qualified pursuant to the laws of the United States or any state or territory thereof, or a certificate as a marine engineer issued by the United States Coast Guard and have had 1 year's experience in the City of New York in the operation and maintenance of stationary high pressure boiler plants under the supervision of a licensed high pressure boiler operating engineer within the 7 years immediately preceding the date of his application; provided that the applicant shall have filed with his application his own signed statement that he is the person named in said certificate together with the supporting signed statements by 3 licensed high pressure boiler operating engineers employed in the City of New York at the time of making of such signed statements; or
6. Have had direct supervision, care, operation and maintenance of a steam generating plant of a governmental building, having boilers of 150 or more hp., for a period of 5 years immediately preceding the date of his application and have had in addition 1 year's experience on high pressure boilers under the direct supervision of a licensed high-pressure boiler operating engineer in the City of New York, within the 7 years immediately preceding the date of his application; or

7. Have successfully completed as a registered apprentice an approved training program recognized by New York State apprenticeship council of at least 2 years and have had at least 3 years experience in the City of New York in the operation and maintenance of high pressure boilers under the supervision of a licensed high pressure boiler operating engineer within the 7 years immediately preceding the date of his application.

Definition: For the purpose of this license a high pressure boiler is a boiler that carries a pressure of more than 15 pounds of steam per square inch and is rated in excess of 10 hp., or if such boiler produces hot water at a pressure of 160 psi or at a temperature over 250°F.

Scope of Examination: This examination will consist of a written test of the multiple-choice type and a practical-oral test. The questions will be based on the operation, maintenance and repair of steam boilers, engines, pumps, turbines and their appurtenances. 70% is required in each test. Candidate must pass both tests of this examination in order to qualify for a license.

Only those candidates who obtain a passing grade in the written test will be given the practical-oral test.

Fee: The $85 fee is to be paid at the time of filing the application for the examination.

A candidate will be permitted to take 3 practical-oral tests on the basis of having passed 1 written test, provided the application for an additional practical-oral test is filed not later than 2 years from the date of the written test. The fee of $95 for each additional practical-oral test shall be paid at the time of filing the application.

NOTE: These requirements are for an applicant to take an examination for a license. Persons who are successful in passing the examination must file an application for the license with the Department of Buildings.

TECHNIQUES OF STUDY AND TEST-TAKING

Although a thorough knowledge of the subject matter is the most important factor in succeeding on your exam, the following suggestions could raise your score substantially. These few pointers will give you the strategy employed on tests by those who are most successful in this not-so-mysterious art. It's really quite simple. Do things right . . . right from the beginning. Make these successful methods a habit. Then you'll get the greatest dividends from the time you invest in this book.

PREPARING FOR THE EXAM

1. *Budget your time.* Set aside definite hours each day for concentrated study. Keep to your schedule.

2. *Study alone.* You will concentrate better when you work by yourself. Keep a list of questions you cannot answer and points you are unsure of to talk over with a friend who is preparing for the same exam. Plan to exchange ideas at a joint review session just before the test.

3. *Eliminate distractions.* Disturbances caused by family and neighbor activities (telephone calls, chit-chat, TV programs, etc.) work to your disadvantage. Study in a quiet, private room.

4. *Use the library.* Most colleges and universities have excellent library facilities. Some institutions have special libraries for the various subject areas: physics library, education library, psychology library, etc. Take full advantage of such valuable facilities. The library is free from those distractions that may inhibit your home study. Moreover, research in your subject area is more convenient in a library since it can provide more study material than you have at home.

5. *Answer all the questions in this book.* Don't be satisfied merely with the correct answer to each question. Do additional research on the other choices which are given. You will broaden your background and be more adequately prepared for the "real" exam. It's quite possible that a question on the exam which you are going to take may require you to be familiar with the other choices.

6. *Get the "feel" of the exam.* The sample questions which this book contains will give you that "feel" since they are virtually the same as those you will find on the test.

7. *Take the Sample Tests as "real" tests.* With this attitude, you will derive greater benefit. Put yourself under strict examination conditions. Tolerate no interruptions while you are taking the sample tests. Work steadily. Do not spend too much time on any one question. If a question seems too difficult go to the next one. If time permits, go back to the omitted question.

8. *Tailor your study to the subject matter. Skim or scan.* Don't study everything in the same manner. Obviously, certain areas are more important than others.

9. *Organize yourself.* Make sure that your notes are in good order—valuable time is unnecessarily consumed when you can't find quickly what you are looking for.

10. *Keep physically fit.* You cannot retain information well when you are uncomfortable, headachy, or tense. Physical health promotes mental efficiency.

HOW TO TAKE AN EXAM

1. *Get to the Examination Room about Ten Minutes Ahead of Time*. You'll get a better start when you are accustomed to the room. If the room is too cold, or too warm, or not well ventilated, call these conditions to the attention of the person in charge.

2. *Make Sure that You Read the Instructions Carefully*. In many cases, testtakers lose credits because they misread some important point in the given directions—example: the *incorrect* choice instead of the *correct* choice.

3. *Be Confident*. Statistics conclusively show that high scores are more likely when you are prepared. It is important to know that you are not expected to answer every question correctly. The questions usually have a range of difficulty and differentiate between several levels of skill.

4. *Skip Hard Questions and Go Back Later*. It is a good idea to make a mark on the question sheet next to all questions you cannot answer easily, and to go back to those questions later. First answer the questions you are sure about. Do not panic if you cannot answer a question. Go on and answer the questions you know. Usually the easier questions are presented at the beginning of the exam and the questions become gradually more difficult.

If you do skip ahead on the exam, be sure to skip ahead also on your answer sheet. A good technique is periodically to check the number of the question on the answer sheet with the number of the question on the test. You should do this every time you decide to skip a question. If you fail to skip the corresponding answer blank for that question, all of your following answers will be wrong.

Each student is stronger in some areas than in others. No one is expected to know all the answers. Do not waste time agonizing over a difficult question because it may keep you from getting to other questions that you can answer correctly.

5. *Guess If You Are Not Sure*. No penalty is given for guessing when these exams are scored. Therefore, it is better to guess than to omit an answer.

6. *Mark the Answer Sheet Clearly*. When you take the examination, you will mark your answers to the multiple-choice questions on a separate answer sheet that will be given to you at the test center. If you have not worked with an answer sheet before, it is in your best interest to become familiar with the procedures involved. Remember, knowing the correct answer is not enough! If you do not mark the sheet correctly, so that it can be machine-scored, you will not get credit for your answers!

In addition to marking answers on the separate answer sheet, you will be asked to give your name and other information, including your social security number. As a precaution bring along your social security number for identification purposes.

Read the directions carefully and follow them exactly. If they ask you to print your name in the boxes provided, write only one letter in each box. If your name is longer than the number of boxes provided, omit the letters that do not fit. Remember, you are writing for a machine; it does not have judgment. It can only record the pencil marks you make on the answer sheet.

Use the answer sheet to record all your answers to questions. Each question, or item, has four or five answer choices labeled (A), (B), (C), (D), (E). You will be asked to choose the letter that stands for the best answer. Then you will be asked to mark your answer by blackening the appropriate space on your answer sheet. Be sure that each space you choose and blacken with your pencil is *completely* blackened. The machine will "read" your answers in terms of spaces blackened. Make sure that only one answer is clearly blackened. If you erase an answer, erase it completely and mark your new answer clearly. The machine will give credit only for clearly marked answers. It does not pause to decide whether you really meant (B) or (C).

Make sure that the number of the question you are being asked on the question sheet corresponds to the number of the question you are answering on the answer sheet. It is a good idea to check the numbers of questions and answers frequently. If you decide to skip a question, but fail to skip the corresponding answer blank for that question, all your answers after that will be wrong.

7. *Read Each Question Carefully.* The exam questions are not designed to trick you through misleading or ambiguous alternative choices. On the other hand, they are not all direct questions of factual information. Some are designed to elicit responses that reveal your ability to reason, or to interpret a fact or idea. It's up to you to read each question carefully, so you know what is being asked. The exam authors have tried to make the questions clear. Do not go astray looking for hidden meanings.

8. *Don't Answer Too Fast.* The multiple-choice questions which you will meet are not superficial exercises. They are designed to test not only your memory but also your under-standing and insight. Do not place too much emphasis on speed. The time element is a factor, but it is not all-important. Accuracy should not be sacrificed for speed.

9. *Materials and Conduct at the Test Center.* You need to bring with you to the test center your Admission Form, your social security number, and several No. 2 pencils. Arrive on time as you may not be admitted after testing has begun. Instructions for taking the tests will be read to you by the test supervisor and time will be called when the test is over. If you have questions, you may ask them of the supervisor. Do not give or receive assistance while taking the exams. If you do, you will be asked to turn in all test materials and told to leave the room. You will not be permitted to return and your tests will not be scored.

PART TWO

Sample Practice
Examinations

ANSWER SHEET FOR
SAMPLE PRACTICE EXAMINATION 1

1 Ⓐ Ⓑ Ⓒ Ⓓ	17 Ⓐ Ⓑ Ⓒ Ⓓ	33 Ⓐ Ⓑ Ⓒ Ⓓ	49 Ⓐ Ⓑ Ⓒ Ⓓ	65 Ⓐ Ⓑ Ⓒ Ⓓ
2 Ⓐ Ⓑ Ⓒ Ⓓ	18 Ⓐ Ⓑ Ⓒ Ⓓ	34 Ⓐ Ⓑ Ⓒ Ⓓ	50 Ⓐ Ⓑ Ⓒ Ⓓ	66 Ⓐ Ⓑ Ⓒ Ⓓ
3 Ⓐ Ⓑ Ⓒ Ⓓ	19 Ⓐ Ⓑ Ⓒ Ⓓ	35 Ⓐ Ⓑ Ⓒ Ⓓ	51 Ⓐ Ⓑ Ⓒ Ⓓ	67 Ⓐ Ⓑ Ⓒ Ⓓ
4 Ⓐ Ⓑ Ⓒ Ⓓ	20 Ⓐ Ⓑ Ⓒ Ⓓ	36 Ⓐ Ⓑ Ⓒ Ⓓ	52 Ⓐ Ⓑ Ⓒ Ⓓ	68 Ⓐ Ⓑ Ⓒ Ⓓ
5 Ⓐ Ⓑ Ⓒ Ⓓ	21 Ⓐ Ⓑ Ⓒ Ⓓ	37 Ⓐ Ⓑ Ⓒ Ⓓ	53 Ⓐ Ⓑ Ⓒ Ⓓ	69 Ⓐ Ⓑ Ⓒ Ⓓ
6 Ⓐ Ⓑ Ⓒ Ⓓ	22 Ⓐ Ⓑ Ⓒ Ⓓ	38 Ⓐ Ⓑ Ⓒ Ⓓ	54 Ⓐ Ⓑ Ⓒ Ⓓ	70 Ⓐ Ⓑ Ⓒ Ⓓ
7 Ⓐ Ⓑ Ⓒ Ⓓ	23 Ⓐ Ⓑ Ⓒ Ⓓ	39 Ⓐ Ⓑ Ⓒ Ⓓ	55 Ⓐ Ⓑ Ⓒ Ⓓ	71 Ⓐ Ⓑ Ⓒ Ⓓ
8 Ⓐ Ⓑ Ⓒ Ⓓ	24 Ⓐ Ⓑ Ⓒ Ⓓ	40 Ⓐ Ⓑ Ⓒ Ⓓ	56 Ⓐ Ⓑ Ⓒ Ⓓ	72 Ⓐ Ⓑ Ⓒ Ⓓ
9 Ⓐ Ⓑ Ⓒ Ⓓ	25 Ⓐ Ⓑ Ⓒ Ⓓ	41 Ⓐ Ⓑ Ⓒ Ⓓ	57 Ⓐ Ⓑ Ⓒ Ⓓ	73 Ⓐ Ⓑ Ⓒ Ⓓ
10 Ⓐ Ⓑ Ⓒ Ⓓ	26 Ⓐ Ⓑ Ⓒ Ⓓ	42 Ⓐ Ⓑ Ⓒ Ⓓ	58 Ⓐ Ⓑ Ⓒ Ⓓ	74 Ⓐ Ⓑ Ⓒ Ⓓ
11 Ⓐ Ⓑ Ⓒ Ⓓ	27 Ⓐ Ⓑ Ⓒ Ⓓ	43 Ⓐ Ⓑ Ⓒ Ⓓ	59 Ⓐ Ⓑ Ⓒ Ⓓ	75 Ⓐ Ⓑ Ⓒ Ⓓ
12 Ⓐ Ⓑ Ⓒ Ⓓ	28 Ⓐ Ⓑ Ⓒ Ⓓ	44 Ⓐ Ⓑ Ⓒ Ⓓ	60 Ⓐ Ⓑ Ⓒ Ⓓ	76 Ⓐ Ⓑ Ⓒ Ⓓ
13 Ⓐ Ⓑ Ⓒ Ⓓ	29 Ⓐ Ⓑ Ⓒ Ⓓ	45 Ⓐ Ⓑ Ⓒ Ⓓ	61 Ⓐ Ⓑ Ⓒ Ⓓ	77 Ⓐ Ⓑ Ⓒ Ⓓ
14 Ⓐ Ⓑ Ⓒ Ⓓ	30 Ⓐ Ⓑ Ⓒ Ⓓ	46 Ⓐ Ⓑ Ⓒ Ⓓ	62 Ⓐ Ⓑ Ⓒ Ⓓ	78 Ⓐ Ⓑ Ⓒ Ⓓ
15 Ⓐ Ⓑ Ⓒ Ⓓ	31 Ⓐ Ⓑ Ⓒ Ⓓ	47 Ⓐ Ⓑ Ⓒ Ⓓ	63 Ⓐ Ⓑ Ⓒ Ⓓ	79 Ⓐ Ⓑ Ⓒ Ⓓ
16 Ⓐ Ⓑ Ⓒ Ⓓ	32 Ⓐ Ⓑ Ⓒ Ⓓ	48 Ⓐ Ⓑ Ⓒ Ⓓ	64 Ⓐ Ⓑ Ⓒ Ⓓ	80 Ⓐ Ⓑ Ⓒ Ⓓ

ANSWER SHEET FOR
SAMPLE PRACTICE EXAMINATION 2

1 Ⓐ Ⓑ Ⓒ Ⓓ   15 Ⓐ Ⓑ Ⓒ Ⓓ   29 Ⓐ Ⓑ Ⓒ Ⓓ   43 Ⓐ Ⓑ Ⓒ Ⓓ   57 Ⓐ Ⓑ Ⓒ Ⓓ

2 Ⓐ Ⓑ Ⓒ Ⓓ   16 Ⓐ Ⓑ Ⓒ Ⓓ   30 Ⓐ Ⓑ Ⓒ Ⓓ   44 Ⓐ Ⓑ Ⓒ Ⓓ   58 Ⓐ Ⓑ Ⓒ Ⓓ

3 Ⓐ Ⓑ Ⓒ Ⓓ   17 Ⓐ Ⓑ Ⓒ Ⓓ   31 Ⓐ Ⓑ Ⓒ Ⓓ   45 Ⓐ Ⓑ Ⓒ Ⓓ   59 Ⓐ Ⓑ Ⓒ Ⓓ

4 Ⓐ Ⓑ Ⓒ Ⓓ   18 Ⓐ Ⓑ Ⓒ Ⓓ   32 Ⓐ Ⓑ Ⓒ Ⓓ   46 Ⓐ Ⓑ Ⓒ Ⓓ   60 Ⓐ Ⓑ Ⓒ Ⓓ

5 Ⓐ Ⓑ Ⓒ Ⓓ   19 Ⓐ Ⓑ Ⓒ Ⓓ   33 Ⓐ Ⓑ Ⓒ Ⓓ   47 Ⓐ Ⓑ Ⓒ Ⓓ   61 Ⓐ Ⓑ Ⓒ Ⓓ

6 Ⓐ Ⓑ Ⓒ Ⓓ   20 Ⓐ Ⓑ Ⓒ Ⓓ   34 Ⓐ Ⓑ Ⓒ Ⓓ   48 Ⓐ Ⓑ Ⓒ Ⓓ   62 Ⓐ Ⓑ Ⓒ Ⓓ

7 Ⓐ Ⓑ Ⓒ Ⓓ   21 Ⓐ Ⓑ Ⓒ Ⓓ   35 Ⓐ Ⓑ Ⓒ Ⓓ   49 Ⓐ Ⓑ Ⓒ Ⓓ   63 Ⓐ Ⓑ Ⓒ Ⓓ

8 Ⓐ Ⓑ Ⓒ Ⓓ   22 Ⓐ Ⓑ Ⓒ Ⓓ   36 Ⓐ Ⓑ Ⓒ Ⓓ   50 Ⓐ Ⓑ Ⓒ Ⓓ   64 Ⓐ Ⓑ Ⓒ Ⓓ

9 Ⓐ Ⓑ Ⓒ Ⓓ   23 Ⓐ Ⓑ Ⓒ Ⓓ   37 Ⓐ Ⓑ Ⓒ Ⓓ   51 Ⓐ Ⓑ Ⓒ Ⓓ   65 Ⓐ Ⓑ Ⓒ Ⓓ

10 Ⓐ Ⓑ Ⓒ Ⓓ   24 Ⓐ Ⓑ Ⓒ Ⓓ   38 Ⓐ Ⓑ Ⓒ Ⓓ   52 Ⓐ Ⓑ Ⓒ Ⓓ   66 Ⓐ Ⓑ Ⓒ Ⓓ

11 Ⓐ Ⓑ Ⓒ Ⓓ   25 Ⓐ Ⓑ Ⓒ Ⓓ   39 Ⓐ Ⓑ Ⓒ Ⓓ   53 Ⓐ Ⓑ Ⓒ Ⓓ   67 Ⓐ Ⓑ Ⓒ Ⓓ

12 Ⓐ Ⓑ Ⓒ Ⓓ   26 Ⓐ Ⓑ Ⓒ Ⓓ   40 Ⓐ Ⓑ Ⓒ Ⓓ   54 Ⓐ Ⓑ Ⓒ Ⓓ   68 Ⓐ Ⓑ Ⓒ Ⓓ

13 Ⓐ Ⓑ Ⓒ Ⓓ   27 Ⓐ Ⓑ Ⓒ Ⓓ   41 Ⓐ Ⓑ Ⓒ Ⓓ   55 Ⓐ Ⓑ Ⓒ Ⓓ   69 Ⓐ Ⓑ Ⓒ Ⓓ

14 Ⓐ Ⓑ Ⓒ Ⓓ   28 Ⓐ Ⓑ Ⓒ Ⓓ   42 Ⓐ Ⓑ Ⓒ Ⓓ   56 Ⓐ Ⓑ Ⓒ Ⓓ   70 Ⓐ Ⓑ Ⓒ Ⓓ

ANSWER SHEET FOR
SAMPLE PRACTICE EXAMINATION 3

1 Ⓐ Ⓑ Ⓒ Ⓓ	17 Ⓐ Ⓑ Ⓒ Ⓓ	33 Ⓐ Ⓑ Ⓒ Ⓓ	49 Ⓐ Ⓑ Ⓒ Ⓓ	65 Ⓐ Ⓑ Ⓒ Ⓓ
2 Ⓐ Ⓑ Ⓒ Ⓓ	18 Ⓐ Ⓑ Ⓒ Ⓓ	34 Ⓐ Ⓑ Ⓒ Ⓓ	50 Ⓐ Ⓑ Ⓒ Ⓓ	66 Ⓐ Ⓑ Ⓒ Ⓓ
3 Ⓐ Ⓑ Ⓒ Ⓓ	19 Ⓐ Ⓑ Ⓒ Ⓓ	35 Ⓐ Ⓑ Ⓒ Ⓓ	51 Ⓐ Ⓑ Ⓒ Ⓓ	67 Ⓐ Ⓑ Ⓒ Ⓓ
4 Ⓐ Ⓑ Ⓒ Ⓓ	20 Ⓐ Ⓑ Ⓒ Ⓓ	36 Ⓐ Ⓑ Ⓒ Ⓓ	52 Ⓐ Ⓑ Ⓒ Ⓓ	68 Ⓐ Ⓑ Ⓒ Ⓓ
5 Ⓐ Ⓑ Ⓒ Ⓓ	21 Ⓐ Ⓑ Ⓒ Ⓓ	37 Ⓐ Ⓑ Ⓒ Ⓓ	53 Ⓐ Ⓑ Ⓒ Ⓓ	69 Ⓐ Ⓑ Ⓒ Ⓓ
6 Ⓐ Ⓑ Ⓒ Ⓓ	22 Ⓐ Ⓑ Ⓒ Ⓓ	38 Ⓐ Ⓑ Ⓒ Ⓓ	54 Ⓐ Ⓑ Ⓒ Ⓓ	70 Ⓐ Ⓑ Ⓒ Ⓓ
7 Ⓐ Ⓑ Ⓒ Ⓓ	23 Ⓐ Ⓑ Ⓒ Ⓓ	39 Ⓐ Ⓑ Ⓒ Ⓓ	55 Ⓐ Ⓑ Ⓒ Ⓓ	71 Ⓐ Ⓑ Ⓒ Ⓓ
8 Ⓐ Ⓑ Ⓒ Ⓓ	24 Ⓐ Ⓑ Ⓒ Ⓓ	40 Ⓐ Ⓑ Ⓒ Ⓓ	56 Ⓐ Ⓑ Ⓒ Ⓓ	72 Ⓐ Ⓑ Ⓒ Ⓓ
9 Ⓐ Ⓑ Ⓒ Ⓓ	25 Ⓐ Ⓑ Ⓒ Ⓓ	41 Ⓐ Ⓑ Ⓒ Ⓓ	57 Ⓐ Ⓑ Ⓒ Ⓓ	73 Ⓐ Ⓑ Ⓒ Ⓓ
10 Ⓐ Ⓑ Ⓒ Ⓓ	26 Ⓐ Ⓑ Ⓒ Ⓓ	42 Ⓐ Ⓑ Ⓒ Ⓓ	58 Ⓐ Ⓑ Ⓒ Ⓓ	74 Ⓐ Ⓑ Ⓒ Ⓓ
11 Ⓐ Ⓑ Ⓒ Ⓓ	27 Ⓐ Ⓑ Ⓒ Ⓓ	43 Ⓐ Ⓑ Ⓒ Ⓓ	59 Ⓐ Ⓑ Ⓒ Ⓓ	75 Ⓐ Ⓑ Ⓒ Ⓓ
12 Ⓐ Ⓑ Ⓒ Ⓓ	28 Ⓐ Ⓑ Ⓒ Ⓓ	44 Ⓐ Ⓑ Ⓒ Ⓓ	60 Ⓐ Ⓑ Ⓒ Ⓓ	76 Ⓐ Ⓑ Ⓒ Ⓓ
13 Ⓐ Ⓑ Ⓒ Ⓓ	29 Ⓐ Ⓑ Ⓒ Ⓓ	45 Ⓐ Ⓑ Ⓒ Ⓓ	61 Ⓐ Ⓑ Ⓒ Ⓓ	77 Ⓐ Ⓑ Ⓒ Ⓓ
14 Ⓐ Ⓑ Ⓒ Ⓓ	30 Ⓐ Ⓑ Ⓒ Ⓓ	46 Ⓐ Ⓑ Ⓒ Ⓓ	62 Ⓐ Ⓑ Ⓒ Ⓓ	78 Ⓐ Ⓑ Ⓒ Ⓓ
15 Ⓐ Ⓑ Ⓒ Ⓓ	31 Ⓐ Ⓑ Ⓒ Ⓓ	47 Ⓐ Ⓑ Ⓒ Ⓓ	63 Ⓐ Ⓑ Ⓒ Ⓓ	79 Ⓐ Ⓑ Ⓒ Ⓓ
16 Ⓐ Ⓑ Ⓒ Ⓓ	32 Ⓐ Ⓑ Ⓒ Ⓓ	48 Ⓐ Ⓑ Ⓒ Ⓓ	64 Ⓐ Ⓑ Ⓒ Ⓓ	80 Ⓐ Ⓑ Ⓒ Ⓓ

ANSWER SHEET FOR
SAMPLE PRACTICE EXAMINATION 4

1 Ⓐ Ⓑ Ⓒ Ⓓ   17 Ⓐ Ⓑ Ⓒ Ⓓ   33 Ⓐ Ⓑ Ⓒ Ⓓ   49 Ⓐ Ⓑ Ⓒ Ⓓ   65 Ⓐ Ⓑ Ⓒ Ⓓ

2 Ⓐ Ⓑ Ⓒ Ⓓ   18 Ⓐ Ⓑ Ⓒ Ⓓ   34 Ⓐ Ⓑ Ⓒ Ⓓ   50 Ⓐ Ⓑ Ⓒ Ⓓ   66 Ⓐ Ⓑ Ⓒ Ⓓ

3 Ⓐ Ⓑ Ⓒ Ⓓ   19 Ⓐ Ⓑ Ⓒ Ⓓ   35 Ⓐ Ⓑ Ⓒ Ⓓ   51 Ⓐ Ⓑ Ⓒ Ⓓ   67 Ⓐ Ⓑ Ⓒ Ⓓ

4 Ⓐ Ⓑ Ⓒ Ⓓ   20 Ⓐ Ⓑ Ⓒ Ⓓ   36 Ⓐ Ⓑ Ⓒ Ⓓ   52 Ⓐ Ⓑ Ⓒ Ⓓ   68 Ⓐ Ⓑ Ⓒ Ⓓ

5 Ⓐ Ⓑ Ⓒ Ⓓ   21 Ⓐ Ⓑ Ⓒ Ⓓ   37 Ⓐ Ⓑ Ⓒ Ⓓ   53 Ⓐ Ⓑ Ⓒ Ⓓ   69 Ⓐ Ⓑ Ⓒ Ⓓ

6 Ⓐ Ⓑ Ⓒ Ⓓ   22 Ⓐ Ⓑ Ⓒ Ⓓ   38 Ⓐ Ⓑ Ⓒ Ⓓ   54 Ⓐ Ⓑ Ⓒ Ⓓ   70 Ⓐ Ⓑ Ⓒ Ⓓ

7 Ⓐ Ⓑ Ⓒ Ⓓ   23 Ⓐ Ⓑ Ⓒ Ⓓ   39 Ⓐ Ⓑ Ⓒ Ⓓ   55 Ⓐ Ⓑ Ⓒ Ⓓ   71 Ⓐ Ⓑ Ⓒ Ⓓ

8 Ⓐ Ⓑ Ⓒ Ⓓ   24 Ⓐ Ⓑ Ⓒ Ⓓ   40 Ⓐ Ⓑ Ⓒ Ⓓ   56 Ⓐ Ⓑ Ⓒ Ⓓ   72 Ⓐ Ⓑ Ⓒ Ⓓ

9 Ⓐ Ⓑ Ⓒ Ⓓ   25 Ⓐ Ⓑ Ⓒ Ⓓ   41 Ⓐ Ⓑ Ⓒ Ⓓ   57 Ⓐ Ⓑ Ⓒ Ⓓ   73 Ⓐ Ⓑ Ⓒ Ⓓ

10 Ⓐ Ⓑ Ⓒ Ⓓ   26 Ⓐ Ⓑ Ⓒ Ⓓ   42 Ⓐ Ⓑ Ⓒ Ⓓ   58 Ⓐ Ⓑ Ⓒ Ⓓ   74 Ⓐ Ⓑ Ⓒ Ⓓ

11 Ⓐ Ⓑ Ⓒ Ⓓ   27 Ⓐ Ⓑ Ⓒ Ⓓ   43 Ⓐ Ⓑ Ⓒ Ⓓ   59 Ⓐ Ⓑ Ⓒ Ⓓ   75 Ⓐ Ⓑ Ⓒ Ⓓ

12 Ⓐ Ⓑ Ⓒ Ⓓ   28 Ⓐ Ⓑ Ⓒ Ⓓ   44 Ⓐ Ⓑ Ⓒ Ⓓ   60 Ⓐ Ⓑ Ⓒ Ⓓ   76 Ⓐ Ⓑ Ⓒ Ⓓ

13 Ⓐ Ⓑ Ⓒ Ⓓ   29 Ⓐ Ⓑ Ⓒ Ⓓ   45 Ⓐ Ⓑ Ⓒ Ⓓ   61 Ⓐ Ⓑ Ⓒ Ⓓ   77 Ⓐ Ⓑ Ⓒ Ⓓ

14 Ⓐ Ⓑ Ⓒ Ⓓ   30 Ⓐ Ⓑ Ⓒ Ⓓ   46 Ⓐ Ⓑ Ⓒ Ⓓ   62 Ⓐ Ⓑ Ⓒ Ⓓ   78 Ⓐ Ⓑ Ⓒ Ⓓ

15 Ⓐ Ⓑ Ⓒ Ⓓ   31 Ⓐ Ⓑ Ⓒ Ⓓ   47 Ⓐ Ⓑ Ⓒ Ⓓ   63 Ⓐ Ⓑ Ⓒ Ⓓ   79 Ⓐ Ⓑ Ⓒ Ⓓ

16 Ⓐ Ⓑ Ⓒ Ⓓ   32 Ⓐ Ⓑ Ⓒ Ⓓ   48 Ⓐ Ⓑ Ⓒ Ⓓ   64 Ⓐ Ⓑ Ⓒ Ⓓ   80 Ⓐ Ⓑ Ⓒ Ⓓ

ANSWER SHEET FOR
SAMPLE PRACTICE EXAMINATION 5

1 Ⓐ Ⓑ Ⓒ Ⓓ	15 Ⓐ Ⓑ Ⓒ Ⓓ	29 Ⓐ Ⓑ Ⓒ Ⓓ	43 Ⓐ Ⓑ Ⓒ Ⓓ	57 Ⓐ Ⓑ Ⓒ Ⓓ
2 Ⓐ Ⓑ Ⓒ Ⓓ	16 Ⓐ Ⓑ Ⓒ Ⓓ	30 Ⓐ Ⓑ Ⓒ Ⓓ	44 Ⓐ Ⓑ Ⓒ Ⓓ	58 Ⓐ Ⓑ Ⓒ Ⓓ
3 Ⓐ Ⓑ Ⓒ Ⓓ	17 Ⓐ Ⓑ Ⓒ Ⓓ	31 Ⓐ Ⓑ Ⓒ Ⓓ	45 Ⓐ Ⓑ Ⓒ Ⓓ	59 Ⓐ Ⓑ Ⓒ Ⓓ
4 Ⓐ Ⓑ Ⓒ Ⓓ	18 Ⓐ Ⓑ Ⓒ Ⓓ	32 Ⓐ Ⓑ Ⓒ Ⓓ	46 Ⓐ Ⓑ Ⓒ Ⓓ	60 Ⓐ Ⓑ Ⓒ Ⓓ
5 Ⓐ Ⓑ Ⓒ Ⓓ	19 Ⓐ Ⓑ Ⓒ Ⓓ	33 Ⓐ Ⓑ Ⓒ Ⓓ	47 Ⓐ Ⓑ Ⓒ Ⓓ	61 Ⓐ Ⓑ Ⓒ Ⓓ
6 Ⓐ Ⓑ Ⓒ Ⓓ	20 Ⓐ Ⓑ Ⓒ Ⓓ	34 Ⓐ Ⓑ Ⓒ Ⓓ	48 Ⓐ Ⓑ Ⓒ Ⓓ	62 Ⓐ Ⓑ Ⓒ Ⓓ
7 Ⓐ Ⓑ Ⓒ Ⓓ	21 Ⓐ Ⓑ Ⓒ Ⓓ	35 Ⓐ Ⓑ Ⓒ Ⓓ	49 Ⓐ Ⓑ Ⓒ Ⓓ	63 Ⓐ Ⓑ Ⓒ Ⓓ
8 Ⓐ Ⓑ Ⓒ Ⓓ	22 Ⓐ Ⓑ Ⓒ Ⓓ	36 Ⓐ Ⓑ Ⓒ Ⓓ	50 Ⓐ Ⓑ Ⓒ Ⓓ	64 Ⓐ Ⓑ Ⓒ Ⓓ
9 Ⓐ Ⓑ Ⓒ Ⓓ	23 Ⓐ Ⓑ Ⓒ Ⓓ	37 Ⓐ Ⓑ Ⓒ Ⓓ	51 Ⓐ Ⓑ Ⓒ Ⓓ	65 Ⓐ Ⓑ Ⓒ Ⓓ
10 Ⓐ Ⓑ Ⓒ Ⓓ	24 Ⓐ Ⓑ Ⓒ Ⓓ	38 Ⓐ Ⓑ Ⓒ Ⓓ	52 Ⓐ Ⓑ Ⓒ Ⓓ	66 Ⓐ Ⓑ Ⓒ Ⓓ
11 Ⓐ Ⓑ Ⓒ Ⓓ	25 Ⓐ Ⓑ Ⓒ Ⓓ	39 Ⓐ Ⓑ Ⓒ Ⓓ	53 Ⓐ Ⓑ Ⓒ Ⓓ	67 Ⓐ Ⓑ Ⓒ Ⓓ
12 Ⓐ Ⓑ Ⓒ Ⓓ	26 Ⓐ Ⓑ Ⓒ Ⓓ	40 Ⓐ Ⓑ Ⓒ Ⓓ	54 Ⓐ Ⓑ Ⓒ Ⓓ	68 Ⓐ Ⓑ Ⓒ Ⓓ
13 Ⓐ Ⓑ Ⓒ Ⓓ	27 Ⓐ Ⓑ Ⓒ Ⓓ	41 Ⓐ Ⓑ Ⓒ Ⓓ	55 Ⓐ Ⓑ Ⓒ Ⓓ	69 Ⓐ Ⓑ Ⓒ Ⓓ
14 Ⓐ Ⓑ Ⓒ Ⓓ	28 Ⓐ Ⓑ Ⓒ Ⓓ	42 Ⓐ Ⓑ Ⓒ Ⓓ	56 Ⓐ Ⓑ Ⓒ Ⓓ	70 Ⓐ Ⓑ Ⓒ Ⓓ

ANSWER SHEET FOR
SAMPLE PRACTICE EXAMINATION 6

1 Ⓐ Ⓑ Ⓒ Ⓓ	11 Ⓐ Ⓑ Ⓒ Ⓓ	21 Ⓐ Ⓑ Ⓒ Ⓓ	31 Ⓐ Ⓑ Ⓒ Ⓓ	41 Ⓐ Ⓑ Ⓒ Ⓓ
2 Ⓐ Ⓑ Ⓒ Ⓓ	12 Ⓐ Ⓑ Ⓒ Ⓓ	22 Ⓐ Ⓑ Ⓒ Ⓓ	32 Ⓐ Ⓑ Ⓒ Ⓓ	42 Ⓐ Ⓑ Ⓒ Ⓓ
3 Ⓐ Ⓑ Ⓒ Ⓓ	13 Ⓐ Ⓑ Ⓒ Ⓓ	23 Ⓐ Ⓑ Ⓒ Ⓓ	33 Ⓐ Ⓑ Ⓒ Ⓓ	43 Ⓐ Ⓑ Ⓒ Ⓓ
4 Ⓐ Ⓑ Ⓒ Ⓓ	14 Ⓐ Ⓑ Ⓒ Ⓓ	24 Ⓐ Ⓑ Ⓒ Ⓓ	34 Ⓐ Ⓑ Ⓒ Ⓓ	44 Ⓐ Ⓑ Ⓒ Ⓓ
5 Ⓐ Ⓑ Ⓒ Ⓓ	15 Ⓐ Ⓑ Ⓒ Ⓓ	25 Ⓐ Ⓑ Ⓒ Ⓓ	35 Ⓐ Ⓑ Ⓒ Ⓓ	45 Ⓐ Ⓑ Ⓒ Ⓓ
6 Ⓐ Ⓑ Ⓒ Ⓓ	16 Ⓐ Ⓑ Ⓒ Ⓓ	26 Ⓐ Ⓑ Ⓒ Ⓓ	36 Ⓐ Ⓑ Ⓒ Ⓓ	46 Ⓐ Ⓑ Ⓒ Ⓓ
7 Ⓐ Ⓑ Ⓒ Ⓓ	17 Ⓐ Ⓑ Ⓒ Ⓓ	27 Ⓐ Ⓑ Ⓒ Ⓓ	37 Ⓐ Ⓑ Ⓒ Ⓓ	47 Ⓐ Ⓑ Ⓒ Ⓓ
8 Ⓐ Ⓑ Ⓒ Ⓓ	18 Ⓐ Ⓑ Ⓒ Ⓓ	28 Ⓐ Ⓑ Ⓒ Ⓓ	38 Ⓐ Ⓑ Ⓒ Ⓓ	48 Ⓐ Ⓑ Ⓒ Ⓓ
9 Ⓐ Ⓑ Ⓒ Ⓓ	19 Ⓐ Ⓑ Ⓒ Ⓓ	29 Ⓐ Ⓑ Ⓒ Ⓓ	39 Ⓐ Ⓑ Ⓒ Ⓓ	49 Ⓐ Ⓑ Ⓒ Ⓓ
10 Ⓐ Ⓑ Ⓒ Ⓓ	20 Ⓐ Ⓑ Ⓒ Ⓓ	30 Ⓐ Ⓑ Ⓒ Ⓓ	40 Ⓐ Ⓑ Ⓒ Ⓓ	50 Ⓐ Ⓑ Ⓒ Ⓓ

SAMPLE PRACTICE EXAMINATION 7

1 Ⓐ Ⓑ Ⓒ Ⓓ	11 Ⓐ Ⓑ Ⓒ Ⓓ	21 Ⓐ Ⓑ Ⓒ Ⓓ	31 Ⓐ Ⓑ Ⓒ Ⓓ	41 Ⓐ Ⓑ Ⓒ Ⓓ
2 Ⓐ Ⓑ Ⓒ Ⓓ	12 Ⓐ Ⓑ Ⓒ Ⓓ	22 Ⓐ Ⓑ Ⓒ Ⓓ	32 Ⓐ Ⓑ Ⓒ Ⓓ	42 Ⓐ Ⓑ Ⓒ Ⓓ
3 Ⓐ Ⓑ Ⓒ Ⓓ	13 Ⓐ Ⓑ Ⓒ Ⓓ	23 Ⓐ Ⓑ Ⓒ Ⓓ	33 Ⓐ Ⓑ Ⓒ Ⓓ	43 Ⓐ Ⓑ Ⓒ Ⓓ
4 Ⓐ Ⓑ Ⓒ Ⓓ	14 Ⓐ Ⓑ Ⓒ Ⓓ	24 Ⓐ Ⓑ Ⓒ Ⓓ	34 Ⓐ Ⓑ Ⓒ Ⓓ	44 Ⓐ Ⓑ Ⓒ Ⓓ
5 Ⓐ Ⓑ Ⓒ Ⓓ	15 Ⓐ Ⓑ Ⓒ Ⓓ	25 Ⓐ Ⓑ Ⓒ Ⓓ	35 Ⓐ Ⓑ Ⓒ Ⓓ	45 Ⓐ Ⓑ Ⓒ Ⓓ
6 Ⓐ Ⓑ Ⓒ Ⓓ	16 Ⓐ Ⓑ Ⓒ Ⓓ	26 Ⓐ Ⓑ Ⓒ Ⓓ	36 Ⓐ Ⓑ Ⓒ Ⓓ	46 Ⓐ Ⓑ Ⓒ Ⓓ
7 Ⓐ Ⓑ Ⓒ Ⓓ	17 Ⓐ Ⓑ Ⓒ Ⓓ	27 Ⓐ Ⓑ Ⓒ Ⓓ	37 Ⓐ Ⓑ Ⓒ Ⓓ	47 Ⓐ Ⓑ Ⓒ Ⓓ
8 Ⓐ Ⓑ Ⓒ Ⓓ	18 Ⓐ Ⓑ Ⓒ Ⓓ	28 Ⓐ Ⓑ Ⓒ Ⓓ	38 Ⓐ Ⓑ Ⓒ Ⓓ	48 Ⓐ Ⓑ Ⓒ Ⓓ
9 Ⓐ Ⓑ Ⓒ Ⓓ	19 Ⓐ Ⓑ Ⓒ Ⓓ	29 Ⓐ Ⓑ Ⓒ Ⓓ	39 Ⓐ Ⓑ Ⓒ Ⓓ	49 Ⓐ Ⓑ Ⓒ Ⓓ
10 Ⓐ Ⓑ Ⓒ Ⓓ	20 Ⓐ Ⓑ Ⓒ Ⓓ	30 Ⓐ Ⓑ Ⓒ Ⓓ	40 Ⓐ Ⓑ Ⓒ Ⓓ	50 Ⓐ Ⓑ Ⓒ Ⓓ

ANSWER SHEET FOR
SAMPLE PRACTICE EXAMINATION 8

1 Ⓐ Ⓑ Ⓒ Ⓓ   16 Ⓐ Ⓑ Ⓒ Ⓓ   31 Ⓐ Ⓑ Ⓒ Ⓓ   46 Ⓐ Ⓑ Ⓒ Ⓓ   61 Ⓐ Ⓑ Ⓒ Ⓓ

2 Ⓐ Ⓑ Ⓒ Ⓓ   17 Ⓐ Ⓑ Ⓒ Ⓓ   32 Ⓐ Ⓑ Ⓒ Ⓓ   47 Ⓐ Ⓑ Ⓒ Ⓓ   62 Ⓐ Ⓑ Ⓒ Ⓓ

3 Ⓐ Ⓑ Ⓒ Ⓓ   18 Ⓐ Ⓑ Ⓒ Ⓓ   33 Ⓐ Ⓑ Ⓒ Ⓓ   48 Ⓐ Ⓑ Ⓒ Ⓓ   63 Ⓐ Ⓑ Ⓒ Ⓓ

4 Ⓐ Ⓑ Ⓒ Ⓓ   19 Ⓐ Ⓑ Ⓒ Ⓓ   34 Ⓐ Ⓑ Ⓒ Ⓓ   49 Ⓐ Ⓑ Ⓒ Ⓓ   64 Ⓐ Ⓑ Ⓒ Ⓓ

5 Ⓐ Ⓑ Ⓒ Ⓓ   20 Ⓐ Ⓑ Ⓒ Ⓓ   35 Ⓐ Ⓑ Ⓒ Ⓓ   50 Ⓐ Ⓑ Ⓒ Ⓓ   65 Ⓐ Ⓑ Ⓒ Ⓓ

6 Ⓐ Ⓑ Ⓒ Ⓓ   21 Ⓐ Ⓑ Ⓒ Ⓓ   36 Ⓐ Ⓑ Ⓒ Ⓓ   51 Ⓐ Ⓑ Ⓒ Ⓓ   66 Ⓐ Ⓑ Ⓒ Ⓓ

7 Ⓐ Ⓑ Ⓒ Ⓓ   22 Ⓐ Ⓑ Ⓒ Ⓓ   37 Ⓐ Ⓑ Ⓒ Ⓓ   52 Ⓐ Ⓑ Ⓒ Ⓓ   67 Ⓐ Ⓑ Ⓒ Ⓓ

8 Ⓐ Ⓑ Ⓒ Ⓓ   23 Ⓐ Ⓑ Ⓒ Ⓓ   38 Ⓐ Ⓑ Ⓒ Ⓓ   53 Ⓐ Ⓑ Ⓒ Ⓓ   68 Ⓐ Ⓑ Ⓒ Ⓓ

9 Ⓐ Ⓑ Ⓒ Ⓓ   24 Ⓐ Ⓑ Ⓒ Ⓓ   39 Ⓐ Ⓑ Ⓒ Ⓓ   54 Ⓐ Ⓑ Ⓒ Ⓓ   69 Ⓐ Ⓑ Ⓒ Ⓓ

10 Ⓐ Ⓑ Ⓒ Ⓓ   25 Ⓐ Ⓑ Ⓒ Ⓓ   40 Ⓐ Ⓑ Ⓒ Ⓓ   55 Ⓐ Ⓑ Ⓒ Ⓓ   70 Ⓐ Ⓑ Ⓒ Ⓓ

11 Ⓐ Ⓑ Ⓒ Ⓓ   26 Ⓐ Ⓑ Ⓒ Ⓓ   41 Ⓐ Ⓑ Ⓒ Ⓓ   56 Ⓐ Ⓑ Ⓒ Ⓓ   71 Ⓐ Ⓑ Ⓒ Ⓓ

12 Ⓐ Ⓑ Ⓒ Ⓓ   27 Ⓐ Ⓑ Ⓒ Ⓓ   42 Ⓐ Ⓑ Ⓒ Ⓓ   57 Ⓐ Ⓑ Ⓒ Ⓓ   72 Ⓐ Ⓑ Ⓒ Ⓓ

13 Ⓐ Ⓑ Ⓒ Ⓓ   28 Ⓐ Ⓑ Ⓒ Ⓓ   43 Ⓐ Ⓑ Ⓒ Ⓓ   58 Ⓐ Ⓑ Ⓒ Ⓓ   73 Ⓐ Ⓑ Ⓒ Ⓓ

14 Ⓐ Ⓑ Ⓒ Ⓓ   29 Ⓐ Ⓑ Ⓒ Ⓓ   44 Ⓐ Ⓑ Ⓒ Ⓓ   59 Ⓐ Ⓑ Ⓒ Ⓓ   74 Ⓐ Ⓑ Ⓒ Ⓓ

15 Ⓐ Ⓑ Ⓒ Ⓓ   30 Ⓐ Ⓑ Ⓒ Ⓓ   45 Ⓐ Ⓑ Ⓒ Ⓓ   60 Ⓐ Ⓑ Ⓒ Ⓓ   75 Ⓐ Ⓑ Ⓒ Ⓓ

ANSWER SHEET FOR
SAMPLE PRACTICE EXAMINATION 9

1 Ⓐ Ⓑ Ⓒ Ⓓ   11 Ⓐ Ⓑ Ⓒ Ⓓ   21 Ⓐ Ⓑ Ⓒ Ⓓ   31 Ⓐ Ⓑ Ⓒ Ⓓ   41 Ⓐ Ⓑ Ⓒ Ⓓ

2 Ⓐ Ⓑ Ⓒ Ⓓ   12 Ⓐ Ⓑ Ⓒ Ⓓ   22 Ⓐ Ⓑ Ⓒ Ⓓ   32 Ⓐ Ⓑ Ⓒ Ⓓ   42 Ⓐ Ⓑ Ⓒ Ⓓ

3 Ⓐ Ⓑ Ⓒ Ⓓ   13 Ⓐ Ⓑ Ⓒ Ⓓ   23 Ⓐ Ⓑ Ⓒ Ⓓ   33 Ⓐ Ⓑ Ⓒ Ⓓ   43 Ⓐ Ⓑ Ⓒ Ⓓ

4 Ⓐ Ⓑ Ⓒ Ⓓ   14 Ⓐ Ⓑ Ⓒ Ⓓ   24 Ⓐ Ⓑ Ⓒ Ⓓ   34 Ⓐ Ⓑ Ⓒ Ⓓ   44 Ⓐ Ⓑ Ⓒ Ⓓ

5 Ⓐ Ⓑ Ⓒ Ⓓ   15 Ⓐ Ⓑ Ⓒ Ⓓ   25 Ⓐ Ⓑ Ⓒ Ⓓ   35 Ⓐ Ⓑ Ⓒ Ⓓ   45 Ⓐ Ⓑ Ⓒ Ⓓ

6 Ⓐ Ⓑ Ⓒ Ⓓ   16 Ⓐ Ⓑ Ⓒ Ⓓ   26 Ⓐ Ⓑ Ⓒ Ⓓ   36 Ⓐ Ⓑ Ⓒ Ⓓ   46 Ⓐ Ⓑ Ⓒ Ⓓ

7 Ⓐ Ⓑ Ⓒ Ⓓ   17 Ⓐ Ⓑ Ⓒ Ⓓ   27 Ⓐ Ⓑ Ⓒ Ⓓ   37 Ⓐ Ⓑ Ⓒ Ⓓ   47 Ⓐ Ⓑ Ⓒ Ⓓ

8 Ⓐ Ⓑ Ⓒ Ⓓ   18 Ⓐ Ⓑ Ⓒ Ⓓ   28 Ⓐ Ⓑ Ⓒ Ⓓ   38 Ⓐ Ⓑ Ⓒ Ⓓ   48 Ⓐ Ⓑ Ⓒ Ⓓ

9 Ⓐ Ⓑ Ⓒ Ⓓ   19 Ⓐ Ⓑ Ⓒ Ⓓ   29 Ⓐ Ⓑ Ⓒ Ⓓ   39 Ⓐ Ⓑ Ⓒ Ⓓ   49 Ⓐ Ⓑ Ⓒ Ⓓ

10 Ⓐ Ⓑ Ⓒ Ⓓ   20 Ⓐ Ⓑ Ⓒ Ⓓ   30 Ⓐ Ⓑ Ⓒ Ⓓ   40 Ⓐ Ⓑ Ⓒ Ⓓ   50 Ⓐ Ⓑ Ⓒ Ⓓ

ANSWER SHEET FOR
SAMPLE PRACTICE EXAMINATION 11

1 Ⓐ Ⓑ Ⓒ Ⓓ   17 Ⓐ Ⓑ Ⓒ Ⓓ   33 Ⓐ Ⓑ Ⓒ Ⓓ   49 Ⓐ Ⓑ Ⓒ Ⓓ   65 Ⓐ Ⓑ Ⓒ Ⓓ

2 Ⓐ Ⓑ Ⓒ Ⓓ   18 Ⓐ Ⓑ Ⓒ Ⓓ   34 Ⓐ Ⓑ Ⓒ Ⓓ   50 Ⓐ Ⓑ Ⓒ Ⓓ   66 Ⓐ Ⓑ Ⓒ Ⓓ

3 Ⓐ Ⓑ Ⓒ Ⓓ   19 Ⓐ Ⓑ Ⓒ Ⓓ   35 Ⓐ Ⓑ Ⓒ Ⓓ   51 Ⓐ Ⓑ Ⓒ Ⓓ   67 Ⓐ Ⓑ Ⓒ Ⓓ

4 Ⓐ Ⓑ Ⓒ Ⓓ   20 Ⓐ Ⓑ Ⓒ Ⓓ   36 Ⓐ Ⓑ Ⓒ Ⓓ   52 Ⓐ Ⓑ Ⓒ Ⓓ   68 Ⓐ Ⓑ Ⓒ Ⓓ

5 Ⓐ Ⓑ Ⓒ Ⓓ   21 Ⓐ Ⓑ Ⓒ Ⓓ   37 Ⓐ Ⓑ Ⓒ Ⓓ   53 Ⓐ Ⓑ Ⓒ Ⓓ   69 Ⓐ Ⓑ Ⓒ Ⓓ

6 Ⓐ Ⓑ Ⓒ Ⓓ   22 Ⓐ Ⓑ Ⓒ Ⓓ   38 Ⓐ Ⓑ Ⓒ Ⓓ   54 Ⓐ Ⓑ Ⓒ Ⓓ   70 Ⓐ Ⓑ Ⓒ Ⓓ

7 Ⓐ Ⓑ Ⓒ Ⓓ   23 Ⓐ Ⓑ Ⓒ Ⓓ   39 Ⓐ Ⓑ Ⓒ Ⓓ   55 Ⓐ Ⓑ Ⓒ Ⓓ   71 Ⓐ Ⓑ Ⓒ Ⓓ

8 Ⓐ Ⓑ Ⓒ Ⓓ   24 Ⓐ Ⓑ Ⓒ Ⓓ   40 Ⓐ Ⓑ Ⓒ Ⓓ   56 Ⓐ Ⓑ Ⓒ Ⓓ   72 Ⓐ Ⓑ Ⓒ Ⓓ

9 Ⓐ Ⓑ Ⓒ Ⓓ   25 Ⓐ Ⓑ Ⓒ Ⓓ   41 Ⓐ Ⓑ Ⓒ Ⓓ   57 Ⓐ Ⓑ Ⓒ Ⓓ   73 Ⓐ Ⓑ Ⓒ Ⓓ

10 Ⓐ Ⓑ Ⓒ Ⓓ   26 Ⓐ Ⓑ Ⓒ Ⓓ   42 Ⓐ Ⓑ Ⓒ Ⓓ   58 Ⓐ Ⓑ Ⓒ Ⓓ   74 Ⓐ Ⓑ Ⓒ Ⓓ

11 Ⓐ Ⓑ Ⓒ Ⓓ   27 Ⓐ Ⓑ Ⓒ Ⓓ   43 Ⓐ Ⓑ Ⓒ Ⓓ   59 Ⓐ Ⓑ Ⓒ Ⓓ   75 Ⓐ Ⓑ Ⓒ Ⓓ

12 Ⓐ Ⓑ Ⓒ Ⓓ   28 Ⓐ Ⓑ Ⓒ Ⓓ   44 Ⓐ Ⓑ Ⓒ Ⓓ   60 Ⓐ Ⓑ Ⓒ Ⓓ   76 Ⓐ Ⓑ Ⓒ Ⓓ

13 Ⓐ Ⓑ Ⓒ Ⓓ   29 Ⓐ Ⓑ Ⓒ Ⓓ   45 Ⓐ Ⓑ Ⓒ Ⓓ   61 Ⓐ Ⓑ Ⓒ Ⓓ   77 Ⓐ Ⓑ Ⓒ Ⓓ

14 Ⓐ Ⓑ Ⓒ Ⓓ   30 Ⓐ Ⓑ Ⓒ Ⓓ   46 Ⓐ Ⓑ Ⓒ Ⓓ   62 Ⓐ Ⓑ Ⓒ Ⓓ   78 Ⓐ Ⓑ Ⓒ Ⓓ

15 Ⓐ Ⓑ Ⓒ Ⓓ   31 Ⓐ Ⓑ Ⓒ Ⓓ   47 Ⓐ Ⓑ Ⓒ Ⓓ   63 Ⓐ Ⓑ Ⓒ Ⓓ   79 Ⓐ Ⓑ Ⓒ Ⓓ

16 Ⓐ Ⓑ Ⓒ Ⓓ   32 Ⓐ Ⓑ Ⓒ Ⓓ   48 Ⓐ Ⓑ Ⓒ Ⓓ   64 Ⓐ Ⓑ Ⓒ Ⓓ   80 Ⓐ Ⓑ Ⓒ Ⓓ

ANSWER SHEET FOR
SAMPLE PRACTICE EXAMINATION 12

1 Ⓐ Ⓑ Ⓒ Ⓓ	17 Ⓐ Ⓑ Ⓒ Ⓓ	33 Ⓐ Ⓑ Ⓒ Ⓓ	49 Ⓐ Ⓑ Ⓒ Ⓓ	65 Ⓐ Ⓑ Ⓒ Ⓓ
2 Ⓐ Ⓑ Ⓒ Ⓓ	18 Ⓐ Ⓑ Ⓒ Ⓓ	34 Ⓐ Ⓑ Ⓒ Ⓓ	50 Ⓐ Ⓑ Ⓒ Ⓓ	66 Ⓐ Ⓑ Ⓒ Ⓓ
3 Ⓐ Ⓑ Ⓒ Ⓓ	19 Ⓐ Ⓑ Ⓒ Ⓓ	35 Ⓐ Ⓑ Ⓒ Ⓓ	51 Ⓐ Ⓑ Ⓒ Ⓓ	67 Ⓐ Ⓑ Ⓒ Ⓓ
4 Ⓐ Ⓑ Ⓒ Ⓓ	20 Ⓐ Ⓑ Ⓒ Ⓓ	36 Ⓐ Ⓑ Ⓒ Ⓓ	52 Ⓐ Ⓑ Ⓒ Ⓓ	68 Ⓐ Ⓑ Ⓒ Ⓓ
5 Ⓐ Ⓑ Ⓒ Ⓓ	21 Ⓐ Ⓑ Ⓒ Ⓓ	37 Ⓐ Ⓑ Ⓒ Ⓓ	53 Ⓐ Ⓑ Ⓒ Ⓓ	69 Ⓐ Ⓑ Ⓒ Ⓓ
6 Ⓐ Ⓑ Ⓒ Ⓓ	22 Ⓐ Ⓑ Ⓒ Ⓓ	38 Ⓐ Ⓑ Ⓒ Ⓓ	54 Ⓐ Ⓑ Ⓒ Ⓓ	70 Ⓐ Ⓑ Ⓒ Ⓓ
7 Ⓐ Ⓑ Ⓒ Ⓓ	23 Ⓐ Ⓑ Ⓒ Ⓓ	39 Ⓐ Ⓑ Ⓒ Ⓓ	55 Ⓐ Ⓑ Ⓒ Ⓓ	71 Ⓐ Ⓑ Ⓒ Ⓓ
8 Ⓐ Ⓑ Ⓒ Ⓓ	24 Ⓐ Ⓑ Ⓒ Ⓓ	40 Ⓐ Ⓑ Ⓒ Ⓓ	56 Ⓐ Ⓑ Ⓒ Ⓓ	72 Ⓐ Ⓑ Ⓒ Ⓓ
9 Ⓐ Ⓑ Ⓒ Ⓓ	25 Ⓐ Ⓑ Ⓒ Ⓓ	41 Ⓐ Ⓑ Ⓒ Ⓓ	57 Ⓐ Ⓑ Ⓒ Ⓓ	73 Ⓐ Ⓑ Ⓒ Ⓓ
10 Ⓐ Ⓑ Ⓒ Ⓓ	26 Ⓐ Ⓑ Ⓒ Ⓓ	42 Ⓐ Ⓑ Ⓒ Ⓓ	58 Ⓐ Ⓑ Ⓒ Ⓓ	74 Ⓐ Ⓑ Ⓒ Ⓓ
11 Ⓐ Ⓑ Ⓒ Ⓓ	27 Ⓐ Ⓑ Ⓒ Ⓓ	43 Ⓐ Ⓑ Ⓒ Ⓓ	59 Ⓐ Ⓑ Ⓒ Ⓓ	75 Ⓐ Ⓑ Ⓒ Ⓓ
12 Ⓐ Ⓑ Ⓒ Ⓓ	28 Ⓐ Ⓑ Ⓒ Ⓓ	44 Ⓐ Ⓑ Ⓒ Ⓓ	60 Ⓐ Ⓑ Ⓒ Ⓓ	76 Ⓐ Ⓑ Ⓒ Ⓓ
13 Ⓐ Ⓑ Ⓒ Ⓓ	29 Ⓐ Ⓑ Ⓒ Ⓓ	45 Ⓐ Ⓑ Ⓒ Ⓓ	61 Ⓐ Ⓑ Ⓒ Ⓓ	77 Ⓐ Ⓑ Ⓒ Ⓓ
14 Ⓐ Ⓑ Ⓒ Ⓓ	30 Ⓐ Ⓑ Ⓒ Ⓓ	46 Ⓐ Ⓑ Ⓒ Ⓓ	62 Ⓐ Ⓑ Ⓒ Ⓓ	78 Ⓐ Ⓑ Ⓒ Ⓓ
15 Ⓐ Ⓑ Ⓒ Ⓓ	31 Ⓐ Ⓑ Ⓒ Ⓓ	47 Ⓐ Ⓑ Ⓒ Ⓓ	63 Ⓐ Ⓑ Ⓒ Ⓓ	79 Ⓐ Ⓑ Ⓒ Ⓓ
16 Ⓐ Ⓑ Ⓒ Ⓓ	32 Ⓐ Ⓑ Ⓒ Ⓓ	48 Ⓐ Ⓑ Ⓒ Ⓓ	64 Ⓐ Ⓑ Ⓒ Ⓓ	80 Ⓐ Ⓑ Ⓒ Ⓓ

SAMPLE PRACTICE EXAMINATION 1

STATIONARY ENGINEER

High Pressure Boiler Operating Engineer

DIRECTIONS: Each question has four suggested answers, lettered A, B, C, and D. Decide which one is the best answer, locate the question number on the sample answer sheet, and with a soft pencil darken the area that corresponds to the answer you have selected. All sample answer sheets follow page 22.

The time allowed for the entire examination is 3½ hours.

1. The bottom blowdown on a boiler is used to

   (A) remove mud drum water impurities
   (B) increase boiler priming
   (C) reduce steam pressure in the header
   (D) increase the boiler water level.

2. The term "spalling" refers to a boiler

   (A) flue gas content
   (B) soot blower
   (C) combustion chamber
   (D) mud leg.

3. The wrench that would normally be used on hexagonally shaped screwed valves and fittings is the

   (A) adjustable pipe wrench
   (B) tappet wrench
   (C) band wrench
   (D) open-end wrench.

4. The designated size of a boiler tube is generally based upon its

   (A) internal diameter
   (B) external diameter
   (C) wall thickness
   (D) weight per foot of length.

5. A fusible plug on a boiler is made primarily of

   (A) selenium   (C) zinc
   (B) tin   (D) iron.

6. The range of pH values for boiler feedwater is normally

   (A) 1 to 2   (C) 9 to 10
   (B) 4 to 6   (D) 12 to 15.

7. Boiler horsepower is defined as the evaporation of

   (A) 900 lbs. of water from and at 212°F.
   (B) 400 lbs. of water from and at 212°F.

(C) 345 lbs. of water from and at 212°F.

(D) 34.5 lbs. of water from and at 212°F.

8. A low pressure air-atomizing oil burner has an operating air pressure range of

(A) 25 to 35 lbs.
(B) 16 to 20 lbs.
(C) 6 to 10 lbs.
(D) 1 to 2 lbs.

9. A superheater is installed in a Sterling boiler mainly for the purpose of raising the temperature of the

(A) secondary air
(B) steam leaving the steam drum
(C) boiler feedwater
(D) primary air.

10. The function of a counterflow economizer in a power plant is to

(A) use flue gases to heat feedwater
(B) raise flue gas temperature
(C) recirculate exhaust steam
(D) pre-heat combustion air.

11. A fire due to spontaneous combustion would most easily occur in a pile of

(A) asbestos sheathing
(B) loose lumber
(C) oil drums
(D) oily waste rags.

12. A "damper regulator," used for combustion control, is operated by

(A) steam pressure
(B) the water column
(C) the boiler pump
(D) a pitot tube.

13. The packing of an expansion joint in a firebrick wall of a combustion chamber is generally made of

(A) silica
(B) brick cement
(C) silicon carbide
(D) asbestos.

14. An open-ended steam pipe, called a steam lance, is usually used on a boiler to

(A) remove soot
(B) bleed the steam header
(C) clean the mud drum
(D) clean chimneys.

15. A high vacuum reading on the fuel oil gauge would indicate

(A) an empty oil tank
(B) high oil temperature
(C) a clogged strainer
(D) worn pump gears.

16. The one of the following boilers that is classified as an internally fired boiler is the

(A) cross - drum straight tube boiler
(B) vertical tubular boiler
(C) Sterling boiler
(D) cross - drum horizontal box-header boiler.

17. Try-cocks are used on a boiler primarily to

(A) check the gauge glass reading
(B) release the steam pressure
(C) drain the water column
(D) blow down the gauge glass.

18. Scale deposits on the tubes and shell of a high pressure boiler are undesirable because the deposits cause

(A) protrusions or roughness
(B) suction
(C) foaming
(D) concentrates.

19. The function of a radiation pyrometer is to measure

(A) boiler water height

(B) boiler pressure
(C) furnace temperature
(D) boiler drum stresses.

20. An engine indicator is generally used to measure

(A) steam temperature
(B) heat losses
(C) errors in gauge readings
(D) steam cylinder pressures.

21. A goose-neck is installed in the line connecting a steam gauge to a boiler to

(A) maintain constant steam flow
(B) prevent steam knocking
(C) maintain steam pressure
(D) protect the gauge element.

22. A boiler steam gauge should have a range of at least

(A) one half the working steam pressure
(B) the working steam pressure
(C) 1½ times the maximum allowable working pressure
(D) twice the maximum allowable pressure.

23. A disconnected steam pressure gauge is usually calibrated with

(A) an Orsat instrument
(B) an air pump
(C) a tuyeres
(D) a dead-weight tester.

24. The recommended size joint for repairing firebrick wall is most nearly

(A) 1/64"   (C) 1/4"
(B) 1/16"   (D) 1/2".

25. The acidity of boiler water is usually determined by a

(A) Rockwell test
(B) soap hardness test
(C) paper test
(D) alkalinity test.

26. Electrostatic precipitators are used in power plants to

(A) remove fly ash from flue gases
(B) measure smoke conditions
(C) collect boiler impurities
(D) disperse minerals in feedwater.

27. Fly ash from the flue gases in a power plant is collected by a

(A) soot blower
(B) gas separator
(C) stack regulator
(D) mechanical separator.

28. The installation of four new split packing rings in a stuffing box requires that the joints of the packing rings be placed

(A) 180° apart   (C) 60° apart
(B) 90° apart   (D) 30° apart.

29. In power plants, boiler feedwater is chemically treated in order to

(A) prevent scale formation
(B) increase water foaming
(C) increase oxygen formation
(D) increase the temperature of the water.

30. The soot in a fire tube boiler generally settles on the

(A) bridgewall
(B) inside tube surface
(C) combustion chamber sides
(D) outside tube surface.

31. The one of the following classifications of fuel oil strainers that is generally not used with the heavier industrial fuel oils is a

(A) wire mesh strainer
(B) metallic disc strainer
(C) filter cloth strainer
(D) perforated metal cylinder strainer.

32. The temperature of the fuel oil leaving a pre-heater is controlled by

    (A) a potentiometer
    (B) a relay
    (C) a low water cut-off
    (D) an aquastat.

33. A pneumatic tool is generally powered by

    (A) natural gas
    (B) steam
    (C) a battery
    (D) air.

34. In New York City the maximum steam pressure permitted in the steam coils used for heating the oil in a submerged oil storage tank is most nearly

    (A) 40 psi   (C) 25 psi
    (B) 35 psi   (D) 10 psi.

35. The water pressure used in a hydrostatic test on a boiler is generally

    (A) 4 times the maximum working pressure
    (B) 2 times the maximum working pressure
    (C) 1-½ times the maximum working pressure
    (D) the same as maximum working pressure.

36. The one of the following valves that should be used in a steam line to throttle the flow is the

    (A) plug valve
    (B) check valve
    (C) gate valve
    (D) globe valve.

37. The CO (carbon monoxide) content in the flue gas from an efficiently fired boiler should be approximately

    (A) 0% to 1%   (C) 8 to 10%
    (B) 4% to 6%   (D) 12 to 13%.

38. The CO_2 (carbon dioxide) percentage in the flue gas of an efficiently fired boiler should be approximately

    (A) 1%   (C) 18%
    (B) 12%   (D) 25%.

39. When the temperature of stack gases rises considerably above the normal operating stack temperature, it generally indicates

    (A) a low boiler water level
    (B) a heavy smoke condition in the stack
    (C) that the boiler is operating efficiently
    (D) that the boiler tubes are dirty.

40. A boiler safety valve is usually set above the maximum working pressure by an amount equal to

    (A) 6% of the maximum working pressure
    (B) 10% of the maximum working pressure
    (C) 12% of the maximum working pressure
    (D) 14% of the maximum working pressure.

41. The one of the following grades of fuel oil that contains the greatest heating value in Btu per gallon is

    (A) #2   (C) #5
    (B) #4   (D) #6.

42. When we say that a fuel oil has a high viscosity, we mean mainly that the fuel oil will

    (A) evaporate easily
    (B) burn without smoke
    (C) flow slowly through pipes
    (D) have a low specific gravity.

43. The type of fuel oil pump generally used with a rotary cup

oil burner system is the

(A) propeller pump
(B) internal pump
(C) centrifugal pump
(D) piston pump.

44. No. 6 fuel oil flowing to a mechanical atomizing burner should be preheated to approximately

(A) 185°F.　　(C) 100°F.
(B) 115°F.　　(D) 80°F.

45. The flame of an industrial rotary cup oil burner should be adjusted so that the flame

(A) has a yellow color with blue spots
(B) strikes all sides of the combustion chamber
(C) has a light brown color
(D) does not strike the rear of the combustion chamber.

46. The location of the oil burner "remote control switch" should generally be

(A) at the boiler room entrance
(B) on the boiler shell
(C) on the oil burner motor
(D) on a wall nearest the boiler.

47. With forced draft the approximate wind box pressure in a single-retort underfeed stoker is normally

(A) 2"　　(C) 7"
(B) 5"　　(D) 9".

48. The pressure over the fire in an oil-fired steam boiler with a balanced-draft system and natural draft is most nearly

(A) + .60"　　(C) -.02"
(B) + .50"　　(D) -.70".

49. Three 2000-gallon tanks with a water content of 10% will yield a weight of most nearly

(A) 600 lbs.　　(C) 5000 lbs.
(B) 6000 lbs.　　(D) 700 lbs.

50. If 30 cu. ft. of air at atmospheric pressure and 60°F. are compressed to a gauge pressure of 80 psi and the temperature rises to 180°F., what is the approximate volume of the compressed air? Assume that the air behaves as a perfect gas and that absolute zero is -460°F.

(A) 5.7 cu. ft.
(B) 7.0 cu. ft.
(C) 14.0 cu. ft.
(D) 16.5 cu. ft.

51. A boiler drum with an inside diameter of 4 ft. is made of 1 in. thick steel having a strength of 50,000 psi. If 0.125 in. of the steel's thickness has been lost due to corrosion, what is the approximate pressure that this boiler can withstand?

(A) 90 psi　　(C) 364 psi
(B) 416 psi　　(D) 1822 psi.

52. What is the additional heat lost due to the emission of 10,000 lbs. of flue gas, if the stack temperature rises from 680°F. to 780°F. while the boiler room temperature remains at 80°F.? Assume that the specific heat of flue gas is 0.25.

(A) 1,000,000 Btu
(B) 375,000 Btu
(C) 250,000 Btu
(D) 125,000 Btu.

53. What is the boiler horsepower of a drum 4 ft. in diameter and 20 ft. long if only 2/3 of the drum's surface contacts the hot gases? Assume that 10 sq. ft. of heating surface are equivalent to a boiler horsepower.

(A) 4.2　　(C) 25.1
(B) 16.7　　(D) 201.1.

54. The monthly cost of firing a certain boiler is $10,000 when its combustion efficiency is 75%. What will the annual savings be if its efficiency is increased to 80%?

   (A) $625 (C) $7,500
   (B) $667 (D) $8,000.

55. Compound gauges are commonly used to read either

   (A) pounds/minute or gallons/minute
   (B) cubic feet/minute or gallons/minute
   (C) vacuum or pressure
   (D) temperature or pressure.

56. The pH of boiler water should be kept at most nearly

   (A) 1 (C) 8
   (B) 5 (D) 10.

57. A major cause of air pollution resulting from the burning of fuel oils is

   (A) sulphur dioxide
   (B) silicon dioxide
   (C) nitrous dioxide
   (D) hydrogen dioxide.

58. The CO_2 percentage in the flue gas of a power plant is indicated by a

   (A) Doppler meter
   (B) Ranarex indicator
   (C) Microtector
   (D) hygrometer.

59. The most likely cause of black smoke exhausting from the chimney of an oil-fired boiler is

   (A) high secondary air flow
   (B) low stack emission
   (C) low oil temperature
   (D) high chimney draft.

60. The diameter of the steam piston in a steam-driven duplex vacuum pump whose dimensions are given as 3 by 2 by 4 is

   (A) 2 (C) 4
   (B) 3 (D) 6.

61 An induced draft fan is generally connected between the

   (A) condenser and the first pass
   (B) stack and the breeching
   (C) feedwater heater and the boiler feed pump
   (D) combustion chamber and fuel oil tanks.

62. The purpose of an air chamber on a reciprocating water pump is to

   (A) maintain a uniform flow
   (B) reduce the amount of steam expansion
   (C) create a pulsating flow
   (D) vary the amount of steam admission.

63. "Flash point" is the temperature at which oil will

   (A) change completely to vapor
   (B) safely fire in a furnace
   (C) flash into flame if a lighted match is passed just above the top of the oil
   (D) burn intermittently when ignited.

64. A "sounding box" would normally be found

   (A) on top of the boiler
   (B) next to a compressed air tank
   (C) in a fuel oil tank
   (D) in a steam condenser.

65. An "intercooler" is generally found on

   (A) a steam pump
   (B) an air compressor
   (C) a steam engine
   (D) a rotary oil pump.

66. The instrument used to measure atmospheric pressure is a

(A) capillary tube
(B) venturi
(C) barometer
(D) calorimeter.

67. The control which starts or stops the operation of the oil burner at a predetermined steam pressure is the

(A) pressuretrol
(B) air flow interlock
(C) transformer
(D) magnetic oil valve.

68. In a closed feedwater heater, the water and the steam

(A) come into direct contact
(B) are kept apart from each other
(C) are under negative pressure
(D) mix and exhaust to the atmosphere.

69. A "knocking" noise in steam lines is generally the result of

(A) superheated steam expansion
(B) high steam pressure
(C) condensation in the line
(D) rapid steam expansion.

70. An electrical component known as a step-up transformer operates by

(A) raising voltage and decreasing amperage
(B) decreasing amperage and raising resistance
(C) raising amperage and decreasing resistance
(D) raising voltage and am-

perage at the same time.

71. A manometer is an instrument that is used to measure

(A) heat radiation
(B) air volume
(C) condensate water level
(D) air pressure.

72. Three 75-gal. per hr. mechanical pressure type oil burners operating together are to burn 150,000 gal. of No. 6 fuel oil. The number of hours they would take to burn this amount of oil is most nearly

(A) 665   (C) 870
(B) 760   (D) 1210.

73. The sum of 10 1/2, 8 3/4, 5 1/2, and 2 1/4 is

(A) 23   (C) 26
(B) 25   (D) 27.

74. A water tank measures 50 ft. long, 16 ft. wide, and 12 ft. high. Assume that water weighs 60 lbs. per cubic ft. and that one gal. of water weights 8 lbs. The number of gallons the tank can hold when it is 1/2 full is

(A) 21,500   (C) 33,410
(B) 28,375   (D) 36,000.

75. Assuming 70 gallons of oil cost $42.00, then 110 gallons of oil at the same rate will cost

(A) $66.00   (C) $96.00
(B) $84.00   (D) $152.00.

Questions 76 to 80 are to be answered in accordance with the information contained in the following paragraph.

Fuel is conserved when a boiler is operating near its most efficient load. The efficiency of a boiler will change as the output varies. Large amounts of air must be used at low ratings and so the heat exchange is inefficient. As the output increases, the efficiency decreases due to an increase in flue gas temperature. Every boiler has an output rate for which its efficiency is highest. For example, in a water-tube boiler the highest efficiency might occur at 120% of rated capacity while in a vertical fire-tube boiler highest efficiency might be at 70% of rated capacity. The type of fuel burned and cleanliness affects the maximum efficiency of the boiler. When a power plant contains a battery of boilers, a sufficient number should be kept in operation so as to maintain the output of individual units near their points of maximum efficiency. One of the boilers in the battery can be used as a regulator to meet the change in demand for steam while the other boilers could still operate at their most efficient rating. Boiler performance is expressed as the number of pounds of steam generated per pound of fuel.

76. According to the above paragraph, the number of pounds of steam generated per pound of fuel is a measure of boiler

(A) size
(B) performance
(C) regulator input
(C) by-pass.

77. According to the above paragraph, the highest efficiency of a vertical fire-tube boiler might occur at

(A) 70% of rated capacity
(B) 80% of water tube capacity
(C) 95% of water tube capacity
(D) 120% of rated capacity.

78. According to the above paragraph, the maximum efficiency of a boiler is affected by the

(A) atmospheric temperature
(B) atmospheric pressure
(C) cleanliness
(D) fire brick material.

79. According to the above paragraph, heat exchanger uses large amounts of air at low

(A) fuel rates
(B) ratings
(C) temperatures
(D) pressures.

80. According to the above paragraph, one boiler in a battery of boilers should be used as a

(A) demand
(B) stand-by
(C) regulator
(D) safety.

Answer Key

1.	A	17.	A	33.	D	49.	C	65.	B
2.	C	18.	A	34.	D	50.	A	66	C
3.	C	19.	C	35.	C	51.	D	67.	A
4.	B	20.	D	36.	D	52.	C	68.	B
5.	B	21.	D	37.	A	53.	B	69.	C
6.	C	22.	C	38.	B	54.	C	70.	A
7.	D	23.	D	39.	D	55.	C	71.	D
8.	D	24.	B	40.	A	56.	C	72.	A
9.	B	25.	D	41.	D	57.	A	73.	D
10.	A	26.	A	42.	C	58.	B	74.	D
11.	D	27.	D	43.	B	59.	C	75.	A
12.	A	28.	B	44.	A	60.	B	76.	B
13.	D	29.	A	45.	D	61.	B	77.	A
14.	A	30.	B	46.	A	62.	A	78.	C
15.	C	31.	C	47.	A	63.	C	79.	B
16.	B	32.	D	48.	C	64.	C	80.	C

SAMPLE PRACTICE
EXAMINATION 2

STATIONARY ENGINEER

High Pressure Boiler Operating Engineer

DIRECTIONS: Each question has four suggested answers, lettered A, B, C, and D. Decide which one is the best answer, locate the question number on the sample answer sheet, and with a soft pencil darken the area that corresponds to the answer you have selected. All sample answer sheets follow page 22.

The time allowed for the entire examination is 3½ hours.

1. In a Scotch-Marine boiler the fusible plug is located

   (A) in the front plate of the boiler
   (B) in the crown sheet
   (C) in the rear wall of the boiler
   (D) on the bottom of the boiler shell so that the contents of the boiler can be drained quickly to a sewer in case of an emergency.

2. Stay bolts are most commonly used in

   (A) fire tube boilers
   (B) water tube boilers
   (C) bent tube boilers
   (D) porcupine boilers.

3. Generally a Sterling boiler has

   (A) a steam drum connected to a mud drum with straight tubes only
   (B) two steam drums connected to two mud drums by means of bent tubes only

   (C) three steam drums connected to a mud drum by means of bent tubes
   (D) two steam drums connected to a mud drum by means of straight tubes only.

4. On an installed stationary water tube boiler it becomes necessary to reset the safety valve. This

   (A) may be done by the plant Chief Engineer under any conditions
   (B) may be done by the watch engineer only in the presence of the plant Chief Engineer
   (C) may be done by the fireman in the presence of watch engineer and the plant Chief Engineer
   (D) may be done by the watch engineer in the presence of the fireman and the City's Boiler Inspector.

5. The soot is removed from boiler tubes by means of a

(A) low pressure flexible air
   lance
(B) steam lance
(C) turbine
(D) slice bar.

6. In New York City the sludge
   and mud from the mud drum is
   usually blown

   (A) directly into the city
      sewer
   (B) into a tank wagon and
      carted away
   (C) into a tank and after a
      cooling down period,
      drained into the city
      sewer
   (D) and pumped by means of a
      centrifugal pump into the
      city sewer.

7. Evaporating 34.5 lbs. of water
   per hr. from a temperature of
   212°F. to steam at atmospheric
   pressure is equivalent to

   (A) a boiler horsepower
   (B) 970 Btu per hr.
   (C) 33,479 Btu per min.
   (D) a microwatt.

8. A stop and check valve is
   usually placed

   (A) in the steam line leaving
      a water tube boiler
   (B) in the blowdown line of a
      water tube boiler
   (C) in the feedwater line at
      the drum
   (D) in the superheater line.

9. The function of the "dry pipe"
   in a water tube boiler is

   (A) that of a collecting pipe
   (B) to dry the wet steam
   (C) to reduce the circulation
      of wet steam
   (D) in the drum to increase
      its steam capacity.

10. A water tube boiler has 70
   tubes, 4 in. outside diameter
   and 20 ft. long. The rated
   boiler hp. based upon the

heating surface of these tubes
is approximately

(A) 450.42   (C) 146.67
(B) 125.34   (D) 552.53.

11. The throttling calorimeter is
   used to determine the

   (A) temperature of superheated
      steam
   (B) amount of moisture in the
      steam
   (C) pressure of the steam at a
      given temperature
   (D) heat of evaporation of
      wet steam.

12. The trycocks on the water
   column are used to

   (A) blow down the water column
   (B) drain the steam drum
   (C) sample the steam and water
      in the drum
   (D) determine the water level
      in the boiler.

13. The number of passes usually
   found in a hand-fired longitudinal drum water tube
   boiler are

   (A) 4   (C) 2
   (B) 3   (D) 1.

14. The tubes of a cross drum
   water tube boiler are usually
   made of

   (A) wrought iron
   (B) copper
   (C) silicon steel
   (D) brass.

15. The baffles in a longitudinal
   drum standard hand-fired Heine
   water tube boiler are placed

   (A) perpendicular to the tubes
   (B) making an angle of 60°
      with the tubes
   (C) parallel with the tubes
   (D) making an angle of 30°
      with the tubes.

16. The grades of fuel oil known as the "Industrial Oils" are

    (A) 1 and 2
    (B) 4, 5, and 6
    (C) 3, 4, and 5
    (D) 2, 3, and 4.

17. A contractor installed a straight run of steel pipe 100 ft. long. If the pipe was installed when the temperature was 40°F., what is the pipe's total length when carrying steam at 340°F.? Assume that the steel's expansion coefficient is .000007 ft. per degree.

    (A) 99.76 ft.   (C) 100.21 ft.
    (B) 99.79 ft.   (D) 100.24 ft.

18. The Btu content per lb. of a high volatile bituminous coal is approximately

    (A) 18,000   (C) 15,000
    (B) 13,000   (D) 12,000.

19. The best variety of bituminous coal for steaming purposes is

    (A) dry bituminous
    (B) long-flaming bituminous
    (C) bituminous coking
    (D) sub-bituminous.

20. The "calorific value" of coal is accurately determined by the use of

    (A) an Orsat instrument
    (B) a calorimeter
    (C) a "Fryite" instrument
    (D) a pyrometer.

21. A steam atomizing fuel oil burner

    (A) makes very little noise when in operation
    (B) is noisy in operation
    (C) when operating most efficiently, uses 3% of the total amount of steam generated

    (D) uses equal amounts of air and steam for atomization.

22. Which one of the following statements is correct?

    (A) The efficiency of an oil-fired water tube boiler does not depend upon the efficiency of the burner.
    (B) Forced draft systems for secondary air must always be used with fuel oil burners.
    (C) The draft in the windbox is always negative.
    (D) Barometric pressure is measured in inches of mercury.

23. Black smoke issuing from a stack may be the result of

    (A) too much secondary air
    (B) an imperfect mixture of air and combustible
    (C) too much primary air
    (D) a temperature high enough to permit complete oxidation of the volatile combustible.

24. Black smoke from a boiler furnace can be reduced by use of

    (A) steam jets
    (B) an injector
    (C) larger openings between grate bars
    (D) an economizer.

25. The draft reading at the base of a chimney, in a steam generating plant that is operating under forced draft, is

    (A) atmospheric pressure
    (B) the minimum negative pressure in the installation
    (D) above atmospheric pressure.

26. The Bourdon gauge is used to measure

    (A) the gal. of oil in the fuel tanks

(B) the steam pressure in lbs. per sq. in.

(C) the quantity of steam generated

(D) the quantity of steam leaving the boiler in cu. ft. per hr.

27. A draft gauge is calibrated in

(A) in. of water
(B) lbs. per sq. ft.
(C) in. of mercury
(D) in. of tri-sodium phosphate.

28. In a balanced draft system the draft gauge reading over the fire is most nearly

(A) +.01 (C) -.15
(B) +.1   (D) -.005.

29. The Ringleman Chart is generally used in a steam generating plant as an indirect means to control the

(A) feedwater fed to the boiler
(B) smoke issuing from the stack
(C) steam supply
(D) boiler feedwater temperature.

30. In a longitudinal drum water tube boiler, the superheater is generally placed

(A) in the first pass
(B) between the first and second passes
(C) in the third pass
(D) in the breeching.

31. Superheaters are usually constructed of

(A) copper
(B) brass
(C) wrought iron
(D) nickel alloy aluminum.

32. Economizers in large steam generating plants are generally used to

(A) increase the temperature of the steam
(B) increase the temperature of the feedwater
(C) reduce the amount of fly-ash
(D) de-superheat the steam.

33. The piston in the water end of a duplex boiler feed pump is

(A) lubricated by means of graphite grease
(B) lubricated by means of cylinder oil
(C) lubricated by means of machine oil
(D) not lubricated by any of the means indicated in (A), (B), or (C).

34. A compound duplex boiler feed pump has

(A) two steam chests
(B) one steam chest
(C) two steam cylinders
(D) four rocker arms.

35. An intercooler is generally used

(A) between the first and second stage of a turbine
(B) between the high and low pressure cylinders of a steam engine
(C) between the first and second stage of a steam driven air compressor.
(D) between the exhaust end of a steam engine and the condenser.

36. Live steam separators are usually located

(A) at the exhaust end of a steam cylinder
(B) between the boiler and the steam engine
(C) on the unloader in a steam-driven air compressor
(D) on the air chamber of a steam-driven boiler feed pump.

37. Exhaust heads

    (A) are used on condensers
    (B) are used on steam engines operating with a condenser
    (C) are used directly on steam boilers to blow steam to atmosphere
    (D) eliminate oil and water from exhaust steam.

38. In order to "soften" feedwater before it enters the boiler, the following chemicals are used:

    (A) sulphuric acid and lime
    (B) magnesium bicarbonate and nitric acid
    (C) lime and soda ash
    (D) calcium sulphate and muriatic acid.

39. Grease usually contains

    (A) soap made from tallow, caustic soda and beeswax
    (B) soap made from lard oil, sulphur and lime
    (C) lard, neat's foot oil and calcium carbonate
    (D) soap, lime and sodium chloride.

40. In a steam generating plant which burns a high ash coal, it would be advisable to use

    (A) an underfeed stoker
    (B) a chain grate stoker
    (C) a sprinkler stoker
    (D) a down draft stoker.

41. The ignition arch is usually used with the following stoker:

    (A) chain grate
    (B) underfeed single retort
    (C) underfeed multiple retort
    (D) sprinkler.

42. One of the greatest difficulties in the satisfactory burning of pulverized coal has resulted from

    (A) the lack of proper metals
    (B) the design of furnaces
    (C) the formation of slag
    (D) the method used for pulverizing the fuel.

43. The heat content of a gallon of No. 6 fuel oil is approximately

    (A) 135,000 Btu
    (B) 143,000 Btu
    (C) 152,000 Btu
    (D) 165,000 Btu.

44. In order to take care of a 10% load increase in a steam generating plant equipped with mechanical atomizing burners, the operator would first

    (A) change the burner tip
    (B) increase the fuel oil pressure
    (C) push the "gun" further into the furnace
    (D) increase the fuel oil temperature.

45. The vaporstat attached to the fan housing of a fully automatic rotary cup burner comes into operation when there is

    (A) flame failure
    (B) too large an increase in primary air
    (C) fan failure
    (D) too large an increase in secondary air.

46. The fuel oil pressure in a mechanical atomizing burner varies from

    (A) 30 to 60 psi
    (B) 40 to 85 psi
    (C) 75 to 100 psi
    (D) 100 to 200 psi.

47. Which one of the following statements is correct?

    (A) The mechanical atomizing oil burner can handle sudden and wide swings in boiler load without appreciable loss in efficiency.

(B) The first cost of mechanical burners is lower than that of steam atomizing burners.

(C) Mechanical burners under natural draft conditions do not require a greater draft at the boiler damper for a given boiler capacity than will steam atomizing burners.

(D) In stationary work, the mechanical burner is capable of handling 3000 lbs. of oil per hr. with a blast pressure of less than 3 in.

48. Which one of the following statements is correct?

(A) Experience has shown that the use of superheated steam with turbines leads to an appreciable loss in economy.

(B) The water consumption of an engine in pounds per indicated horsepower is in no sense a true indication of its absolute efficiency.

(C) The lower the heat consumption of an engine per indicated horsepower, the lower its economy.

(D) An accurate statement can be made as to the savings possible through the use of superheated steam with reciprocating engines.

49. The power output of a given DeLaval impulse turbine depends upon

(A) the steam pressure alone
(B) the number of nozzles in action
(C) the number of vanes
(D) the number of fixed blades.

50. The vanes in a Curtis turbine are usually made of

(A) wrought iron
(B) steel
(C) copper
(D) nickel bronze.

51. A turbine operating condensing is most efficient when operating with a vacuum of

(A) 29 in. of mercury
(B) 27 in. of mercury
(C) 25 in. of mercury
(D) 23 in. of mercury.

52. A stationary compound steam engine has

(A) three cylinders
(B) four cylinders
(C) two cylinders
(D) one cylinder.

Questions 53-55 are based upon the diagram shown below.

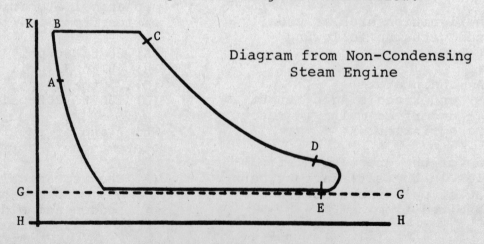

Diagram from Non-Condensing Steam Engine

53. The atmospheric line is

    (A) KH　　　　(C) HH
    (B) GG　　　　(D) CD.

54. The line indicating expansion is

    (A) ABC　　　　(C) AE
    (B) GE　　　　(D) CD.

55. The point of admission is

    (A) C　　　　(C) A
    (B) B　　　　(D) E.

56. In a uniflow engine the exhaust ports are located

    (A) in the head of the piston
    (B) at the center of the cylinder
    (C) at the end of the cylinder
    (D) in the shaft of the piston.

57. The main function of the steam jacket in a steam engine is to

    (A) increase the expansion of the steam in the cylinder
    (B) decrease the amount of steam fed to the cylinder
    (C) reduce initial condensation in the cylinder
    (D) decrease the speed of the engine.

58. The most economical cut-off for a simple steam engine is about

    (A) 1/3 the stroke
    (B) 1/6 the stroke
    (C) 1/2 the stroke
    (D) 1/12 the stroke.

59. In reference to steam engines the phrase "throw of the eccentric" means the

    (A) total distance moved by the steam valve

    (B) angle through which the eccentric must be rotated to cause the steam edge to travel from its central position
    (C) distance from the center of the shaft to the periphery of the eccentric disc.
    (D) distance from the center of the shaft to the center of the eccentric disc.

60. The operating factor which is maintained constant by an inertia type governor is

    (A) speed
    (B) torque
    (C) rate of flow of steam
    (D) rate of exhaust.

61. A compound-wound motor is

    (A) an alternating current machine
    (B) a direct current machine
    (C) a constant torque machine
    (D) a constant speed machine.

62. A large squirrel cage type electric motor

    (A) has a relatively large starting torque per ampere
    (B) draws a small starting current from the line
    (C) has extremely rapid acceleration
    (D) is usually used for constant-speed service with infrequent starting.

63. Flash overs on a commutator are usually caused by

    (A) too light a load on the machine
    (B) a heavy overload
    (C) operating the machine at too low a speed
    (D) operating a 110% of amperage rating.

Questions 64-66 are based upon the problem given below.

Data as Observed

Steam pressure lbs. per sq. in. gauge	151.0
Barometer, in. of mercury	28.5
Temperature of feedwater, F.	161.8
Temperature of furnace, F.	2100.0
Temperature of flue gases, F.	480.0
Temperature of boiler room, F.	60.0
Quality of steam, percent	98.0
Water apparently evaporated, #/hr.	86000.0
Coal as fired, #/hr.	10000.0
Refuse removed from ash pit, #/hr.	1600.0

Coal Analysis, percent of coal as fired

Moisture   8;   Ash   12
               Btu per lb. 11,250

Heat of evaporation (151.0 psi gauge) = 856.8 Btu/#
Heat of liquid (151.0 psi gauge) = 338.2 Btu/#

64. The factor of evaporation is

(A) 1.50   (C) 1.68
(B) 1.35   (D) 1.08.

65. The calorific value of the combustible as fired in Btu/# is most nearly

(A) 14,062   (C) 12,462
(B) 11,250   (D) 13,492.

66. The pounds of water apparently evaporated per pound of combustible as fired is most nearly

(A) 8.60 #water/#combustible
(B) 11.35 #water/#combustible
(C) 10.75 #water/#combustible
(D) 12.75 #water/#combustible.

67. A steam-electric generating plant is set up to develop both 220 volt D.C. and 110 volt D.C. The 220 volt output is rated at 1200 amperes and the 110 volt output is rated at 800 amperes. Under these conditions the 110 volt output is most nearly

(A) 32.2% of the total electrical capacity

(B) 28.2% of the total electrical capacity
(C) 22.2% of the total electrical capacity
(D) 18.1% of the total electrical capacity.

68. A non-condensing engine taking steam at a pressure of 100 lbs. absolute and cutting off at one quarter stroke has a mean effective pressure on the piston of 44.6 psi. If the engine was to exhaust into a condenser against a 26.5 in. vacuum (1.7 pounds absolute), the percentage of power gained would be most nearly

(A) 35.2%   (C) 16.8%
(B) 29.2%   (D) 15.4%.

69. The power required to drive the condensate pump for a turbine installation when operating under the following conditions:
maximum output of main turbine
water rate = 15 lbs. per kw hr.
= 10,000 kw hr.
vacuum = 28 in. of mercury

Given data: Suction head corresponding to 28 in. of mercury = 31 ft.; friction and discharge head = 29 ft.; pump and motor efficiency = 50% is approximately

(A) 12.0 Br. hp.
(B) 11.5 Br. hp.
(C) 9.4 Br. hp.
(D) 8.2 Br. hp.

70. A steam pipe is to carry 5,000 lbs. of steam per hr. at a velocity of 5,000 ft. per min. What is the required nominal inside diameter of this pipe if one pound of the steam has a volume of 5 cubic ft.?

(A) 2 in.　　(C) 16 in.
(B) 4 in.　　(D) 31 in.

Answer Key

1.	B	15.	C	29.	B	43.	C	57.	C
2.	A	16.	B	30.	B	44.	B	58.	A
3.	C	17.	C	31.	C	45.	C	59.	A and D
4.	D	18.	B and C	32.	B	46.	D	60.	A
5.	B	19.	A	33.	D	47.	A	61.	B
6.	C	20.	B	34.	A	48.	B	62.	D
7.	A	21.	B	35.	C	49.	B	63.	B
8.	C	22.	D	36.	B	50.	D	64.	D
9.	A	23.	B	37.	D	51.	A	65.	A
10.	C	24.	A	38.	C	52.	C	66.	C
11.	B	25.	C	39.	A	53.	B	67.	A
12.	D	26.	B	40.	A and B	54.	D	68.	B
13.	B	27.	A	41.	A	55.	C	69.	C
14.	A	28.	D	42.	C	56.	B	70.	B

SAMPLE PRACTICE EXAMINATION 3

STATIONARY ENGINEER

High Pressure Boiler Operating Engineer

DIRECTIONS: Each question has four suggested answers, lettered A, B, C, and D. Decide which one is the best answer, locate the question number on the sample answer sheet, and with a soft pencil darken the area that corresponds to the answer you have selected. All sample answer sheets follow page 22.

The time allowed for the entire examination is 3½ hours.

1. Before a newly installed boiler is placed in operation it is recommended that it be "boiled out" for a period ranging from 36 to 72 hours. A suggested method is to use soda ash and caustic soda. If a steam boiler operator were to use this method properly, then for each 1000 lbs. of water which the boiler holds at the steaming level, he should use most nearly

   (A) 10 lbs. each of soda ash and caustic soda
   (B) 5 lbs. of soda ash and 10 lbs. of caustic soda
   (C) 9 lbs. each of soda ash and caustic soda
   (D) 3 lbs. each of soda ash and caustic soda.

2. With regard to fire-tube boilers, the letters H.R.T. most likely mean

   (A) Highly Regulated Temperature
   (B) Highly Regulated Thermostat
   (C) Horizontal Return Tubular
   (D) Horizontally Rotated Tubes.

3. Baffles are generally arranged in water tube boilers to control and direct the flow of flue gas. In a given boiler the baffles are arranged to direct the flow of flue gas essentially parallel to the tube length. This is commonly referred to as

   (A) cross baffling
   (B) vertical baffling
   (C) angle baffling
   (C) longitudinal baffling.

4. Straight water tubes are sometimes constructed with front and rear "box headers." The front box headers may be riveted directly to the steam drum and hand holes are provided to allow tube inspection, cleaning and replacement. Stay bolts are provided and connect the plate and tube sheet in order to prevent

   (A) bulging under internal pressure

(B) blistering due to internal heat
(C) collapsing under internal vacuum
(D) bagging under internal weight.

5. Straight inclined water tube boilers may be constructed in either of two general designs: longitudinal drum arrangement, or cross drum arrangement. A feature of the cross drum arrangement is that this type of construction generally results in a

(A) loss of head room
(B) saving of head room
(C) loss of combustion of furnace volume
(D) loss of economy for large capacities.

6. A boiler which has been in operation for approximately one year without any "outage" period is to be taken off the line for internal inspection. (This boiler is one of five in a battery with the other four remaining in operation.) This one boiler has to be properly cooled and emptied. Before any man is permitted to enter the drum the correct safety procedure requires that

(A) he be equipped with safety light and clothing
(B) he be provided with a safety line
(C) the drum be ventilated and the safety valve be gagged
(D) the main steam, blowdown and feedwater "stops" be closed, chained and tagged.

7. In keeping with the most recent local law a high pressure steam boiler is one that has a capacity of more than 10 hp. to generate steam and carries a pressure of more than

(A) 10 psi   (C) 25 psi
(B) 15 psi   (D) 75 psi.

8. In a particular plant all the boilers have straight inclined water tubes rolled into sectional headers of serpentine design. The upper bank of tubes are all 1" dia., and the lower bank are all 4" dia. In these boilers the hand holes for the upper tube banks (all hand holes of same size) would most likely permit easy access to

(A) individual tubes
(B) super heater tubes
(C) economizer tubes
(D) a group of tubes.

9. In a particular plant all of the steam boilers are of the Sterling type. In order to maintain the outside surface of the tubes in the boiler in good condition, an operator should use

(A) the hot-lime water treatment procedure
(B) the zeolite system of water treatment
(C) soot blowers such as Diamond or equal
(D) a combination of induced and forced draft.

10. In a given plant the boilers are of the locomotive type. In order to maintain the outside surface of the tubes in good condition while the boiler is in operation, an operator should use

(A) recommended feedwater treatment as directed
(B) Diamond soot blowers of the rotating impulse type
(C) Diamond soot blowers of the telescope retraction type
(D) periodic surface blows.

11. Several of the vertical fire-tube boilers are so designed that the tubes extend above the operating water level. In the continuous and prolonged operation of a boiler of this design at rated load, a possible result would be

(A) an expanded lower tube sheet
(B) repeated burning out of the fusible plug
(C) generation of steam with a very high degree of super heat
(D) leaks at the upper tube sheet due to possible tube overheating.

12. The economizer is the device by means of which feedwater temperature is raised by heat recovery from the flue gas. From a thermal point of view the use of an economizer increases the steam-generator efficiency. The results of tests show that for each 10° to 11°F. increase in feedwater temperature, the efficiency increases approximately

(A) 1%   (C) 7.5%
(B) 5%   (D) 3.5%.

13. Which one of the following statements is correct?

(A) A "pendant" type superheater is always self draining.
(B) The "overdeck" convection type superheater with two headers, both set below the tube elements, is always self draining.
(C) In a radiant type superheater the final steam temperature tends to rise as the load increases.
(D) In a convection type superheater the final steam temperature tends to fall as the load increases.

14. Riveted joints in boiler drums and sheets must be made in accordance with the local code. A triple-riveted butt joint with unequal straps, if used, must have a total of

(A) six rows of rivets
(B) nine rows of rivets
(C) three rows of rivets
(D) five rows of rivets (three for wider strap).

15. The design of a steam generator (boiler) is directed toward the final erection and operation of the unit as a step in the use of heat in the conversion of energy from one form to another. The operation of the steam generator (boiler) should be carried on in such manner as to assure, as a primary consideration, the

(A) conservation of fuel
(B) conservation of materials
(C) reduction of risks to personnel
(D) reduction of waste in the use of steam.

16. An operating steam generator is equipped with three safety valves which are numbered and located as follows: #D-1 on the boiler steam drum, #D-2 on the boiler steam drum, and #S-1 on outlet header of the superheater. These safety valves should be set so that

(A) all three blow at the same time
(B) #S-1 blows first
(C) #D-1 blows first
(D) #D-1 and #D-2 blow at the same pressure ahead of #S-1.

17. A given boiler is one of the bent-tube type with the drum axis parallel to the long axis of the furnace. The design is such that the upper drum is twice the length of the mud drum and with front, half length, side headers to provide for a water-cooled furnace. Feedwater enters the rear of the mud drum and is directed from there

through tubes in the last convection pass to the upper drum. A bridge wall is set across the furnace directly in front of the mud drum. This arrangement for feedwater travel is

(A) desirable because an economizer action is thus developed
(B) desirable because one less connection is made into the upper drum giving a higher ligament co-efficient
(C) desirable because cooler flue gas eliminates the need for induced draft fans
(D) not desirable because relatively cool water is being directed through tubes across which flow the hottest flue gases.

18. With regard to the boiler described in the previous question, the boiler tubes, with respect to size, are arranged as follows: larger tubes in the water walls and first two passes; and smaller tubes, properly nested, in the last pass, arranged with baffles to give cross flow. This arrangement of smaller tubes and cross baffles in the last pass is desirable (although draft loss is increased) because it

(A) gives greater heat absorbing area
(B) is easier to handle and repair smaller diameter tubes
(C) gives better support to the cross baffles
(D) gives lower internal frictional loss (fluid flow).

19. Water-tube boilers fall into two fundamental groups: straight tube type, or bent tube type. The particular advantage, from a maintenance point of view, which favors the straight tube type is that

(A) hanging baffles are more readily supported
(B) individual tubes are more readily replaced
(C) by-pass dampers are more accessible
(D) dusting and access doors may be set on any one of three sides.

20. A straight water tube type boiler has been baffled in such a manner as to hold a relatively constant flue gas velocity through the several passes. Under this condition it can be said that the cross-sectional area of the

(A) first pass is less than that of the second or third pass
(B) second pass is less than that of the first or third pass
(C) third pass is less than that of the first or second pass
(D) three passes are equal.

21. A given boiler is to be taken "off the line" for annual inspection and repairs. In keeping with a good suggested shut down procedure, the boiler may be emptied (drained) when

(A) the setting is cool enough to permit entry into the furnace
(B) the boiler water temperature has dropped to approximately 180°F.
(C) the pressure has fallen to approximately 1/2 of the operating pressure
(D) 48 hours have elapsed after killing the fires.

22. In most power plants the accepted procedure is to blow down the boilers at least once every 24 hours (when blowdown is not continuous or controlled by analysis of the boiler water). The best time for this single blowdown is when the steam rate is at

    (A) 80% of capacity
    (B) 125% of capacity
    (C) 50% of capacity
    (D) its lowest point.

23. Practically all boilers are equipped in such a manner that the gauge glass and water column are provided with blowdown valves. In most plants the operating rules provide that the water column and gauge glass be blown down

    (A) not more than once each two watches (8-hour watches) in under-fired plants
    (B) at least once each watch (8-hour watch)
    (C) at least once each day (24 hours)
    (D) whenever the engineer on watch may think of doing so.

24. A given boiler equipped with a superheater is being placed "on the line" to operate along with a battery of other boilers (200 psi and 100°F. superheat). In accepted operating procedure the drains on the superheater outlet header should remain open until boiler pressure reaches

    (A) 25 psi $\pm$ (C) 180 psi $\pm$
    (B) 50 psi $\pm$ (D) 5 psi.

25. As a Stationary Engineer with an operating department in New York City, you are asked to consider the factors which should be taken into account in the selection of a new steam generator. Of the following, the one which has the least effect upon the selection of a new steam generator (boiler) is the

    (A) type of fuel to be used
    (B) pressure and temperature of steam required
    (C) expected steam load variations
    (D) type of electrical energy to be generated, either A.C. or D.C.

26. Various authorities agree that "spalling" of the refractory lining in the furnace of a steam generator is said to exist when the "refractory cracks and breaks away at the surface." A primary cause of spalling is most likely due to

    (A) uneven heating and cooling within the refractory brick
    (B) continuous overfiring of the boiler
    (C) slag accumulations on furnace walls
    (D) change in fuel from solid to liquid type.

27. For laying-up a boiler for prolonged out-of-service periods, the alternate to the "dry method" is the "wet method." The wet method provides that after the boiler has been completely cleaned and checked, it be filled to the vent valve with deaerated water to which a solution of caustic soda and sodium sulphite has been added. The pH value of this fluid in the boiler should be controlled and maintained in the range of

    (A) 1 to 4   (C) 6 to 11
    (B) 3 to 6   (D) 13 to 14.

28. In a properly operated steam generator the amount of blowdown is adjusted to some per-

centage of the steam flow, as approximated from tests, to keep the boiler-water solids concentration at or below some set limit. If we assume a limit of 850 ppm solids in boiler water, and 30% make up with 110 ppm solids, then for each 1000 lbs. of steam generated the quantity of blowdown should be most nearly

(A) 40 lbs.　　(C) 80 lbs.
(B) 60 lbs.　　(D) 100 lbs.

29. A steam generator is to be laid up for a rather long period of time. Of the several methods recommended for laying-up the boiler the dry method is to be used with the unit closed tightly. In this method the drums and mud boxes are sealed after first placing them in trays of

(A) hot zeolite
(B) unslaked lime
(C) trisodium phosphate
(D) calcium chloride.

30. A given boiler is being fired in a proper manner for warm-up before being placed on the line. This boiler is one of a battery of four, all connected to the same header. This one boiler is equipped with a non-return valve. The purpose of this non-return valve is to

(A) prevent flow of steam from the boiler until boiler pressure is above the header pressure
(B) prevent flow of feedwater from the boiler until boiler pressure is above the feedwater main pressure
(C) prevent flow of air into the boiler as it is being warmed up
(D) prevent flow of steam through the gauge glass in the event it should break.

31. A turbo-generator receives steam at 1200 Btu/lb. and exhausts it at 1050 Btu/lb. Assume there are no losses, and that one kilowatt-hour is equivalent to 3,412 Btu. The steam rate for this turbine is most nearly

(A) 1　　　(C) 23
(B) 3　　　(D) 228.

32. Bottom blowdowns are periodically done on boilers to remove

(A) soot from the gas passes
(B) mud and sediment from the boiler
(C) make-up water for analysis
(D) flue gases for analysis.

33. The number of Btu required to change 10 gallons of water to steam, assuming the water is at 212°F. and atmospheric pressure, is most nearly

(A) none
(B) 2,120 Btu
(C) 9,700 Btu
(D) 80,800 Btu.

34. Boiler drafts are usually measured by draft gauges calibrated in

(A) inches of water
(B) inches of mercury
(C) psi g.
(D) psi a.

35. The Ringleman chart is used to

(A) design ventilating ducts
(B) check the quality of super-heated steam
(C) obtain the heat content of steam
(D) classify the density of smoke.

36. Boiler draft gauges are usually calibrated to read in

(A) psi gauge
(B) psi absolute

(C) inches of mercury
(D) inches of water.

37. Boiler plants are sometimes equipped with Economizers in order to utilize

(A) flue gases for feedwater heating
(B) flue gases for fuel oil preheating
(C) exhaust steam for feedwater heating
(D) exhaust steam for fuel oil heating.

In order to assure the operating engineer in a steam generating plant that he is firing his fuel most efficiently, periodic checks should be made of the combustion recording instruments. These checks are usually made with an Orsat Apparatus. The Orsat Apparatus usually consists of three pipettes, one each for the absorption of CO_2, O_2, and CO. Questions 38-40 relate to the "absorbing reagent" that is contained in each of the pipettes respectively.

38. The absorbing reagent which is used to absorb CO_2 is

(A) potassium pyrogallate
(B) cuprous chloride
(C) potassium hydroxide
(D) hydrogen dioxide.

39. The absorbing reagent is used to absorb CO is

(A) hydrogen dioxide
(B) potassium hydroxide
(C) potassium pyrogallate
(D) cuprous chloride.

40. The absorbing reagent which is used to absorb O_2 is

(A) potassium pyrogallate
(B) potassium hydroxide
(C) cuprous chloride
(D) hydrogen dioxide.

41. Many factors are involved in

the design of a furnace that should cover every possible operating condition. A device commonly used to preserve and prolong the life of furnace walls where furnace temperatures are high due to overfiring is

(A) the ignition-arch
(B) the bridgewall
(C) the water-wall
(D) the back-arch.

42. A poor grade of anthracite coal is properly burned in a furnace. A complete analysis of the flue gas is made. The one constituent which is probably present in this flue gas and which is potentially most corrosive is

(A) H_2O (C) SO_2
(B) CO   (D) CO_2.

43. Many existing steam generators that are coal fired (both hand and mechanical) are equipped with a system of balanced draft. With this system operating properly the over fire draft should be most likely

(A) + 1.02" w.g.
(B) − 1.02" w.g.
(C) + 0.00" w.g.
(D) − 0.02" w.g.

44. An operating boiler is generating superheated steam at 200 #/sq." and total temperature of 420°F. The operating engineer notes that the flue gas temperature in the stack is close to 1000°F. This unit is equipped with both an economizer and an air preheater. Of the following, the one set of conditions which is most likely not responsible for the 1000°F. flue gas temperature is

(A) improper operation of the

soot blower
(B) broken baffles in the boiler
(C) overfiring the boiler at a high rate
(D) a very high percentage of CO_2 in the flue gas.

45. The manner in which fuel oils are treated before firing depends greatly on their viscosity. The most widely accepted standard test method which may be used to determine the viscosity of the various grades of fuel oil is commonly known as the

(A) Saybolt Universal Viscosimeter
(B) Kinematic Universal Viscosimeter
(C) Absolute Universal Viscosimeter
(D) Potential Universal Viscosimeter.

46. A particular boiler is fired by means of straight mechanical pressure type (no return flow) burners using No. 6 oil. The operating pressure range of these burners is such that atomization and operating characteristics become unstable if the pressure at the base of the nozzle falls below

(A) 180 psi   (C) 50 psi
(B) 120 psi   (D) 80 psi.

47. Exhaust steam enters a surface condenser at the rate of 10,000 lbs. per hr. with a heat content of 1050 Btu per lb. The condensate has a heat content of 50 Btu per lb. How many gallons per minute of cooling water must be circulated through this condenser if it enters at 70°F. and leaves at 80°F.? Assume that a gallon of water weighs 8.3 lbs.

(A) 1,000,000 gpm
(B) 16,700 gpm
(C) 2,008 gpm
(D) 1,200 gpm.

48. With regard to the purchase of fuel oils, the "degree API at 60°F." is generally always given in the specification. With regard to this factor, one can always say that the calorific value in Btu/gal. will

(A) increase as the degree API at 60°F. increases
(B) decrease as the degree API at 60°F. increases
(C) not be affected by changes in the degree API at 60°F.
(D) increase as the degree API at 60°F. rises to 25 and thereafter decrease.

49. A development in the mechanical straight pressure type burner design is the use of a double tube gun. One tube is used to feed oil to the tip and the other provides a passage for the unused oil to be returned. In operation with No. 6 fuel oil the important advantage of this design is that it

(A) gives good atomization with a wide variation in capacity
(B) permits the use of the flat flame as against the hollow conical flame
(C) permits the use of a fuel oil piping system with both high and low suction lines
(D) allows the operator to fire fuel oil efficiently at reasonably low temperatures (approximately 120°F.).

50. In order to fire fuel oil efficiently it must be properly atomized immediately before it is mixed with the combustion air. The type of fuel oil

burner that is generally considered most costly to operate
is a

(A) high-pressure air
atomizing burner
(B) steam atomizing burner
(C) low-pressure air
atomizing burner
(D) mechanical atomizing
burner.

Gasket materials, for use in making
up pipe flanges, must be made of
materials which will not be chemically or thermally affected by the
fluid in the pipe. Questions 51
and 52 relate to the recommended
service for given gasket materials.

51. Gaskets made of red rubber are
most generally used for

(A) oil temperatures up to
200°F.
(B) steam or water for temperatures up to 1000°F.
(C) water for temperatures up
to 700°F.
(D) air for temperatures up to
200°F.

52. Gaskets made of corrugated
copper are most generally used
for

(A) steam for temperatures up
to 600°F.
(B) steam or water for temperatures up to 1000°F.
(C) oil for temperatures up to
1000°F.
(D) steam, water or oil for
temperatures up to 1000°F.

53. An industrial heating process
in a given plant uses bled
steam at 30 psi. Accurate and
close temperature control is
required for this process and
the steam trap used to drain
condensate must be of the continuously operating type.
Therefore, the type of steam
trap which most definitely
should not be used is the

(A) combination float and
thermostat
(B) thermostatic
(C) combination inverted
bucket and thermostatic
(D) ball float with external
thermostatic element.

54. A deaerating type feedwater
heater is an essential piece
of equipment in steam-electric
generating stations. This is
especially so if the generated
steam pressures are comparatively high. This type of
feedwater heater removes non-
condensible gas such as air,
free oxygen, CO_2, etc. If
these are permitted to remain
in the feedwater, they tend to

(A) gum up the feedwater
strainers and pump valves
(B) increase filling and
corrosion in the boilers
(C) increase "surging" in the
boiler and "blistering"
in the tubes
(D) destroy the lead metallic
gaskets used in the pipe
flanges.

55. For a given plant you are preparing to purchase valve seats
and disks for the overhaul of
the water end of a duplex
direct-acting steam driven
boiler feedwater pump. The
valve disks for this service
should be made of

(A) hard molded rubber
(B) neoprene
(C) machined cast iron
(D) bronze.

56. Several types of duplex direct-
acting steam driven pumps are
in use for feedwater service
in many generating plants. A
basic difference relates to the
use of the piston type and the
outside packed plunger type
pumps. From the point of view
of the operating and maintenance personnel in a given

plant, the outside packed plunger type is considered more desirable. This opinion is acceptable because

(A) this type of pump is heavier and therefore more enduring
(B) for equal capacity, considerably less steam is used
(C) all packing leakage is external and is a guide for making adjustments
(D) plunger and packing friction is greatly reduced.

57. Pumps of the centrifugal type are generally manufactured in either one of two general patterns: turbine with a diffuser ring, or volute without a diffuser ring. In the volute pattern pumps the discharger chamber acts as a diffuser. The primary purpose for the use of a diffuser ring or chamber is to

(A) convert high velocity to pressure
(B) convert high pressure to velocity
(C) convert the end thrust to a balanced side thrust
(D) bring about a balanced end thrust condition.

58. The recommended procedure for starting a turbine type centrifugal pump handling 100°F. water which has a submerged suction is

(A) prime pump, start pump motor, open discharge valve, and then open suction valve
(B) open suction valve, start pump motor, when pump is up to speed gradually open the discharge valve
(C) open discharge valve, start pump motor, when pump is up to speed gradually open suction valve

(D) open discharge valve, start pump motor, prime pump, and then open suction valve gradually.

59. Some steam generating plants are equipped with injectors as standby equipment for their duplex direct-acting steam pumps. Assume that these injectors are of the single-tube lifting type. When one of these injectors is placed in operation, the steam will first enter

(A) a suction tube
(B) a combining tube
(C) an expanding nozzle
(D) a delivery tube.

60. A duplex direct-acting steam driven pump with outside packed plungers is being put back in operation after being completely overhauled. As part of the overhaul work the steam valves were replaced and set up with greatly reduced lost motion. When this pump is started the probability is that it will operate

(A) at greatly reduced speed
(B) at somewhat increased speed
(C) with a short stroke
(D) with a long stroke.

61. Upon receiving a shipment of thermometers you find that all of the instruments are calibrated with the centigrade instead of the fahrenheit scale. If you had to use one of these thermometers to determine the temperature of boiler feedwater, then a reading of 55°C. would be equal to

(A) 131°F.   (C) 138.6°F.
(B) 99°F.   (D) 62.5°F.

62. The water end of a duplex direct-acting steam driven plunger type boiler feedwater pump (double acting) has a bore of 3" and a stroke of 6".

During the course of a one-hour test run, the speed is maintained constant at 50 rpm. Test data as follows are recorded:

Average water temperature = 160°F.

Total weight of water
pumped   = 1620#

Specific gravity @ 160 F. = 0.98

Wt. of a gal. of water
@ 160 F.   = 8.16#

Volume of one U.S. gal.   = 231
                        cu. in.

Under the above conditions, the slip in this pump is most nearly

  (A) 10%   (C) 3%
  (B) 5%   (D) 1/2%.

63. In addition to the manufacturer's name, address, shop and serial number, the name plate data on a water tube boiler gives the following information:

Number of 4" tubes: 60
Number of 2" tubes: 150
Length of tubes: 25'

Other data, such as total tube heating surface, boiler horsepower, etc., are so worn as to be unreadable. In operation, the boiler generates saturated steam at 125 psi. (The total

heat required to generate a pound of steam under operating conditions is 1040 Btu/#.) Disregarding drums, the total square footage of generating surface is most nearly

  (A) 41,000   (C) 9,100
  (B) 18,100   (D) 3,500.

64. In reference to question statement #63, assume that the average heat transfer rate through the total tube surface is 3000 Btu/sq.'/hr. when operating at full load. Under these conditions, the hourly steam generating rate becomes most nearly

  (A) 26,300 #/hr.
  (B) 10,200 #/hr.
  (C) 52,400 #/hr.
  (D) 118,400 #/hr.

65. For a given ventilating system in a building, it is desired that the c.f.m. handled by the fan in the system be increased by 50%, i.e., from 8000 c.f.m. to 12000 c.f.m. The fan is motor driven by means of a V-belt drive. The present fan speed is 450 rpm. In order to increase the fan output as given above, the new fan speed should be most nearly

  (A) 1010 rpm   (C) 750 rpm
  (B) 875 rpm   (D) 675 rpm.

66. The following operation conditions exist for a given boiler:

Steam Pressure	125 #/sq." gauge
Steam Temperature (Assume steam is delivered saturated and dry.)	353 F.
Feedwater Temperature	190 F.
Total Heat of Steam as delivered (enthalpy)	1,193 Btu/#
Heat of liquid at 190 F.	158 Btu/#
Calorific value of coal as fired	13,000 Btu/#

As a result of the most recent test, the overall boiler efficiency is to be taken as 71%.

In keeping with the above data we may expect that for each pound of coal fired, the pounds of steam generated will be most nearly

(A) 10.91   (C) 8.91
(B) 12.55   (D) 6.81.

67. A given boiler in operation generates 50,000 #/hr. of steam at 200 #/sq." gauge, total temperature of 440°F., and total heat (enthalpy) of 1234.0. Feedwater is delivered to this boiler at a temperature of 190°F., and the heat of the liquid at this temperature is 158 Btu/#. The latent heat of vaporization for water at atmospheric pressure and 212°F. is 970.3 Btu/#. For the above outlined operating conditions the factor of evaporation is most nearly

(A) 1.11   (C) 1.43
(B) 1.27   (D) 0.903.

68. After a good indicator card has been taken for a steam engine, the following equation is used to figure the indicated horsepower:

$$i.h.p. \quad = \quad \frac{PLAN}{K}$$

When N is given in rpm the numerical value of the constant "K" in the above equation is most commonly equal to

(A) 5,250   (C) 1,728
(B) 33,000   (D) 550.

69. A simplex double-acting horizontal reciprocating steam engine is equipped with corliss type valve gear. This valve gear is of the releasing type. In order to get a sharp cut-off

(A) a 600-w lubricant must be used to keep the valve gear free

(B) properly adjusted dashpots must be used
(C) a centrifugal inertia type governor must be used
(D) a single eccentric for control of all four valves must be used.

70. If a copper coil is rotated in a clockwise direction at a uniform speed in a uniform magnetic field, the induced voltage in the coil will be alternating in direction. To obtain Direct Current from this coil when connected to an external circuit, it is common practice to use

(A) two slip rings
(B) one slip ring
(C) a commutator
(D) a balancer coil.

71. Electrical apparatus such as a motor or generator which has been idle for some time in a damp location may have accumulated moisture and is therefore unsafe to put into operation. The apparatus should be dried out thoroughly before being returned to service and periodic insulation resistance measurements taken while in the process of drying out. This is done by using

(A) a growler
(B) an ammeter
(C) a wattmeter
(D) a megger.

72. In reference to a dynamo, brush holders are used to guide and eliminate vibration of brushes, which is a common cause of sparking. The brushes are held against the commutator surface by adjustable springs. For best operation the spring tension should be adjusted so that brushes press against the commutator with a force in lbs. per sq. in. of most nearly

(A) 1.5 to 2.0
(B) 0.5 to 1.0
(C) 2.75 to 3.50
(D) 3.75 to 4.50.

73. A power plant uses direct current generators to deliver electrical energy to a 3-wire balanced system of distribution. Assuming that two balance coils are used but are placed outside of the generator, then connection to the armature winding is usually made through the following number of slip-rings:

(A) 2   (C) 1
(B) 4   (D) 3.

74. A steam engine is equipped with a flyball governor which is belt-driven from a "driving" pulley mounted concentric with the engine shaft. This belt gives motion to a "receiving" pulley which is mounted on a stub shaft which, through bevel gears, gives motion to the governor shaft. Other things remaining the same, if the diameter of the receiving pulley is reduced, this engine would then most likely operate

(A) at a comparatively higher speed
(B) at the speed which existed before
(C) with a reduced cut-off
(D) at a comparatively lower speed.

75. A nine-stage impulse turbine is one in which the stationary vanes between the stages act to

(A) reheat the steam
(B) redirect the steam
(C) re-expand the steam
(D) recompress the steam.

76. A "rateau" or velocity stage is very often made part of a multi-stage impulse turbine.

With respect to a nine-stage impulse turbine the velocity stage is generally the

(A) third stage
(B) ninth stage
(C) first stage
(D) fifth stage.

77. You are asked to assist in setting up and taking indicator cards for a corliss type steam engine. The steam line pressure is 200 psi, and the overall pencil rise of the indicator is 2½". The best "scale of the indicator spring" that should be used, if the allowable pencil rise is limited to 2", is most nearly

(A) 40   (C) 100
(B) 50   (D) 200.

78. The design of a particular turbine is such that the critical speed is above the operating speed. Assume that you are assigned to place this turbine in operation. After warming up both steam header and turbine you slowly begin to bring the unit up to rated speed (3600 rpm). At approximately 1400 rpm the turbine rotor begins to vibrate severely. As the operator you should

(A) increase the speed rapidly and re-open the turbine drains
(B) check the overspeed trip to make sure no one has tampered with it
(C) check the lube oil pressure to make sure it is sufficiently high
(D) decrease the speed quickly and recheck drains, lube oil and warm-up procedure.

79. A simplex reciprocating steam engine is equipped with an inertia type governor and is

coupled to a D.C. generator. In operation the load is dropped from 100% to 50%. As a result, the flywheel tends to

(A) overspeed and the governor bar tends to lag the flywheel
(B) overspeed and the governor bar tends to lead the flywheel
(C) slow down and the governor bar tends to lead the flywheel
(D) slow down and the governor bar tends to lag the flywheel.

80. Upon inspecting the indicator card for a particular steam engine, you note that the atmospheric line is entirely above the exhaust line of the card. As a result you may reasonably assume that this engine was at that time running

(A) non-condensing
(B) overloaded
(C) underloaded
(D) condensing.

Answer Key

1.	D	17.	A	33.	C	49.	A	65.	D
2.	C	18.	A	34.	A	50.	A	66.	C
3.	D	19.	B	35.	D	51.	D	67.	A
4.	A	20.	C	36.	D	52.	A	68.	B
5.	B	21.	A	37.	A	53.	C	69.	B
6.	D	22.	D	38.	C	54.	B	70.	C
7.	B	23.	B	39.	D	55.	D & A	71.	D
8.	D	24.	C	40.	A	56.	C	72.	A
9.	C	25.	D	41.	C	57.	A	73.	B
10.	A	26.	A	42.	C	58.	B	74.	D
11.	D	27.	C	43.	D	59.	C	75.	B
12.	A	28.	A	44.	D	60.	C	76.	C
13.	B	29.	B	45.	A	61.	A	77.	C
14.	A	30.	A	46.	C	62.	A	78.	D
15.	C	31.	C	47.	C	63.	D	79.	A
16.	B	32.	B	48.	B	64.	B	80.	D

SAMPLE PRACTICE EXAMINATION 4

STATIONARY ENGINEER

High Pressure Boiler Operating Engineer

DIRECTIONS: Each question has four suggested answers, lettered A, B, C, and D. Decide which one is the best answer, locate the question number on the sample answer sheet, and with a soft pencil darken the area that corresponds to the answer you have selected. All sample answer sheets follow page 22.

The time allowed for the entire examination is 3½ hours.

1. A Sterling boiler usually has

   (A) straight inclined fire tubes
   (B) serpentine or staggered headers
   (C) horizontal fire-tire baffles
   (D) three steam drums and a mud drum.

2. A steam boiler is blown down in order to

   (A) test the safety valves
   (B) discharge sediment and scale-forming material
   (C) raise the boiler water line
   (D) clean the fire side of tubes.

3. In the average horizontal water tube boiler

   (A) fusible plugs are located in the mud drum
   (B) scale forms on the outside of the tubes
   (C) feed water enters at the front tube header

   (D) the width of the grate determines the spacing "number of tubes wide."

4. Horizontal return tabular boilers are usually constructed so that

   (A) the grate is independent of the boiler
   (B) the tubes are expanded directly into the drum shell
   (C) the tubes are expanded into pressed-steel headers
   (D) vertical baffles are invariably used.

5. The ends of tubes of a steam boiler are usually secured to sheets or headers by

   (A) brazing
   (B) welding
   (C) expanding
   (D) screwed threads.

6. A Scotch-Marine boiler can be identified very easily because it usually

   (A) has fire tubes

(B) has water tubes
(C) is externally fired
(D) requires a brick setting.

7. If a steam boiler is fitted with a superheater coil the superheater safety valve should be set

(A) two lbs. lower than the boiler safety valves
(B) five lbs. lower than the boiler safety valves
(C) two lbs. higher than the boiler safety valves
(D) at the same pressure as the boiler safety valves.

8. In a vertical fire tube boiler operating at 100% of its rated horsepower the sq. ft. of heating surface per boiler horsepower is most nearly

(A) 8   (C) 12
(B) 10   (D) 34.5.

9. The correct water level in a steam boiler is checked by operating the try-cocks and finding that

(A) water only flows out of top try-cocks
(B) steam only flows out of bottom try-cock
(C) steam and water flow out of bottom try-cock
(D) steam and water flow out of middle try-cock.

10. A steam pressure gauge is usually connected to a boiler by means of a loop or pig tail in order to

(A) keep water from entering gauge
(B) prevent steam from entering gauge
(C) take care of pipe expansion
(D) protect the gauge against too high a steam pressure.

11. In order to check the water gauge glass reading of a steam boiler in operation, you should

(A) increase the boiler feed-water pressure
(B) increase the steam pressure
(C) first reduce steam pressure
(D) open the bottom drain valve on the gauge glass.

12. Hard scale which is deposited on the tubes of a water tube boiler

(A) is usually removed by hand operated steam lance through side of boiler
(B) can only be removed by means of a mechanical hammer type cleaner
(C) is easily removed by shutting down boiler and wire-brushing tubes by hand
(D) is usually removed by hydraulic turbine cutters.

13. A water tube boiler has 63 tubes each being 15 ft. long and 4 in. outside diameter. On the basis of tube heating surface, the rated boiler horsepower is most nearly

(A) 145   (C) 200
(B) 75   (D) 100.

14. One boiler horsepower is equivalent to

(A) the evaporation of 34.5 lbs. of water per hr. from 70 F to 212 F.
(B) 33,470 Btu per min.
(C) the evaporation of 34.5 lbs. of water per hr. from a temperature of 212° to steam at same temperature
(D) 970.3 Btu per hr.

15. When filled level, a coal bunker 50 ft. long by 21 ft. wide and 10 ft. high will hold approximately (assume coal occupies 30 cu. ft. per ton)

(A) 300 tons
(B) 350 tons
(C) 530 tons
(D) 600 tons.

16. In judging the quality of boiler feed water with respect to hardness, a fair or moderately hard water is one having most nearly the following grains of incrusting solids per U. S. gallon:

   (A) over 30   (C) 8
   (B) 18   (D) less than 6.

17. Soot which is deposited on the outside of boiler tubes is usually removed by using a

   (A) steam lance
   (B) hydraulic turbine
   (C) low pressure air lance
   (D) chip bar.

18. In a water tube boiler, overheating tubes might be caused by

   (A) carrying too high a water level
   (B) too rapid a flow of water in the tubes
   (C) allowing scale to accumulate on the inside surface of the tubes
   (D) cleaning scale too often from outside surface of the tubes.

19. The "draft" over a fire in a steam boiler is usually expressed as

   (A) negative pressure in inches of air
   (B) positive pressure in inches of air
   (C) negative pressure in inches of water
   (D) none of the above.

20. A compound steam pressure gauge is capable of registering

   (A) the difference between two pressures
   (B) the sum of two line pressures
   (C) either pressure or vacuum

   (D) two different pressures at one time.

21. Suppose that saturated steam at 170 psi/g. is released into the atmosphere and its specific volume changes from 2.7 cubic ft. per lb. to 29.7 cubic ft. per lb. Under these conditions the steam's volume increases by a factor of

   (A) 2.7   (C) 27
   (B) 11   (D) 29.7.

22. Steam with a heat content of 1500 Btu/lb. is supplied to a turbine. Assume that this turbine could exhaust steam in into a condenser at 1100 Btu/lbs. What is the percentage change in the thermal efficiency of this turbine if it exhausts into the atmosphere at 1150 Btu/lb.?

   (A) 11% increase
   (B) 11% decrease
   (C) 13% increase
   (D) 13% decrease.

23. The type of coal having a calorific value of 15,000 to 15,500 Btu per lb. is

   (A) bituminous (mid-continental field)
   (B) anthracite
   (C) high volatile
   (D) sub-bituminous.

24. The type of coal which has the highest amount of "fixed carbon" is

   (A) anthracite
   (B) semi-bituminous
   (C) bituminous
   (D) black lignite.

25. The order in which anthracite coal is classified from the largest steaming size to the smallest is

   (A) pea, barley, buckwheat, rice

(B) pea, buckwheat, rice, barley

(C) pea, rice, barley, buckwheat

(D) buckwheat, pea, barley, rice.

26. The order in which bituminous coal is graded and sized from the largest to the smallest is

(A) slack, nut, lump and run of mine

(B) nut, lump, slack, run of mine

(C) run of mine, slack, lump, nut

(D) run of mine, lump, nut, slack.

27. A report of the "proximate analysis" of anthracite coal "as received" will indicate the percentage by

(A) volume of volatile hatter, fixed carbon, moisture and ash

(B) volume of carbon, hydrogen, oxygen, nitrogen, and sulphur

(C) weight of moisture, fixed carbon, volatile hatter and ash

(D) weight of moisture, hydrogen, sulphur and ash.

28. The principal advantage of using an automatic mechanically operated stoker in a steam power boiler is

(A) the ability of the stoker to meet the sudden changes in load

(B) greater ease in obtaining good economy

(C) the coal is distributed evenly

(D) very little loss of unburned coal with ashes.

29. The standard type of chain-grate stoker is commonly used for burning

(A) small sizes of free-burning bituminous coal

(B) coal having about 10% volatile matter

(C) coal with a low percentage of ash

(D) coking coal.

30. The underfeed type of stoker as used in a steam generating plant

(A) works best with natural draft

(B) operates best with a short fire

(C) depends on a coking arch for smokeless combustion

(D) is essentially a forced draft stoker.

31. The type of stoker best adapted to operating at about 400% of rating is the

(A) chain-grate stoker

(B) overfeed step grate stoker

(C) underfeed stoker

(D) sidefeed step grate stoker.

32. The speed in rpm of the cup of a motor driven horizontal rotary type oil burner is most nearly

(A) 3450   (C) 1800

(B) 2450   (D) 1200.

33. One of the major advantages of the horizontal burner is that

(A) the shape of the fire can be controlled by the shape of the cup

(B) there are comparatively few moving parts

(C) the air nozzle need never be cleaned

(D) the air nozzle controls both the primary and secondary air.

34. When the horizontal rotary burner operates under normal conditions the air nozzle

causes the fire to commence from from the atomizing cup a distance of most nearly

(A) right at the edge of the cup
(B) 1/2 inch
(C) 3/4 inch
(D) several inches.

35. If the fire size of a horizontal rotary burner is reduced so low that it begins to break away from the cup, and this cannot be overcome by closing the primary air damper, then

(A) the air nozzle size should be increased
(B) the air nozzle size should be reduced
(C) the secondary air damper should be opened
(D) the fuel oil pressure should be reduced.

36. Atomization of fuel oil in a horizontal rotary burner is primarily due to

(A) the rotating cup only
(B) the whirling action of the secondary air
(C) the rotating cup and primary air
(D) the pressure of the oil.

37. The calorific value of No. 5 fuel oil in Btu per gallon is most nearly

(A) 148,500   (C) 162,000
(B) 145,500   (D) 152,000.

38. With respect to a mechanical pressure atomizing oil burner, which one of the following statements is true?

(A) A mechanical pressure burner will not operate satisfactorily with natural draft.
(B) A single burner will have a maximum oil-burning

capacity, with forced draft, of about 200 gal. per hr.
(C) This type burner is extremely noisy in operation.
(D) Frequent cleaning of the burner tip and disk is unnecessary.

39. The consumption of a steam for atomization in steam-atomizing burners should be kept below a certain percent of the steam generated. This percent should be most nearly

(A) 5   (C) 6
(B) 2   (D) 8.

40. One advantage of the steam atomizing burner is that

(A) wet steam can be used for atomization
(B) compressed air at 25 lbs. per sq. in. pressure can be used in place of steam
(C) the fuel oil pressure to burner is usually low
(D) dirty oil cannot be used.

41. An injector is used on a higher pressure steam boiler to

(A) blow scum from water surface
(B) mix chemicals with the boiler feed water
(C) feedwater to the boiler drum
(D) suck air from the boiler.

42. If a horizontal compound steam engine has its cylinders so arranged that the two pistons are on the same piston rod, this engine is called a

(A) duplex compound engine
(B) cross compound engine
(C) tandem compound engine
(D) double acting compound engine.

43. In a large uniflow engine the clearance volume, in percent

of cylinder volume, is usually
less than most stationary engines. It is approximately

(A) 1%   (C) 10%
(B) 3%   (D) 15%.

44. Assuming the average height of
an indicated diagram is 1.2
inches and the scale of
spring used is 40, then the
M.E.P. is most nearly

(A) 48 lbs. per sq. in.
(B) 48 lbs. per sq. ft.
(C) 120 lbs. per sq. in.
(D) 40 lbs. per sq. in.

Questions 45-48 are based upon the
diagram shown below:

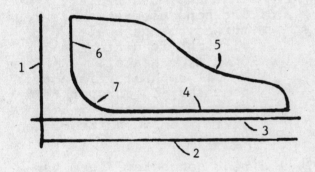

45. The vacuum line is number

(A) 1   (C) 3
(B) 2   (D) 4.

46. The atmosphere line is number

(A) 4   (C) 2
(B) 3   (D) 1.

47. The exhaust line is number

(A) 1   (C) 4
(B) 6   (D) 3.

48. Admission is indicated by line

(A) 7   (C) 4
(B) 5   (D) 6.

49. The horsepower of a double
acting 12 x 35 low speed engine running 100 rpm with mean
effective pressure of 30 lbs./
sq. in. is most nearly (single
acting engine hp = PLAN)
                        33000

(A) 60   (C) 720
(B) 30   (D) 120.

50. The "D" slide valves of a duplex boiler feed pump as a
rule

(A) have a greater stem lap
than exhaust lap when in
mid-travel
(B) are adjusted so that lock
nuts on the valve stem are
always tight
(C) have no lap
(D) should never be stopped on
dead center.

51. If a steam engine valve has
neither lap nor lead, the eccentric should be placed with
respect to the crank

(A) at least 110° in advance
of crank
(B) 90° in advance of crank
(C) 60° in advance of crank
(D) 45° in advance of crank.

52. A foot valve is commonly used
on the submerged end of the
suction pipe of a centrifugal
pump

(A) only when brine is being
pumped
(B) to reduce the load on pump
when starting
(C) as a safety measure in
case the discharge valve
on pump is accidentally
closed while pump is still
running
(D) to keep the pump primed.

53. If the flow of water from a
contrifugal pump should be
stopped by closing the discharge valve with pump still
running,

(A) the pressure will go up
indefinitely
(B) a relief valve will open
(C) the head or pressure will
not go up indefinitely
(D) the pump will overload the
driving motor.

54. Reciprocating pumps are not
usually used

(A) for suction lift without
a priming device
(B) for fire protection equipment
(C) in place of centrifugal
pumps where a higher suc-
tion lift is wanted
(D) for viscous liquids having
a viscosity higher than
500 sec. Saybolt.

55. In regards to a steam driven
duplex boiler feed pump

(A) the suction valves at the
water end are practically
always attached to the
upper plate
(B) the valves at the water
end are practically al-
ways on the flat disc
type
(C) the water piston diameter
is larger than the steam
piston diameter
(D) the water piston diameter
and the steam piston dia-
meter are the same.

56. In a steam power plant operating condensing, the boiler
feed water is heated by exhaust steam passed through an
open type heater. If the
heater is located in the exhaust system between the prime
mover and the condenser, the
heater is commonly known as

(A) a vacuum heater
(B) an auxiliary heater
(C) a pressure heater
(D) a secondary heater.

57. In regards to a closed type
feedwater heater

(A) steam and water come into
direct contact with each
other
(B) the boiler feed pump is
usually placed between the
heater and the boiler
(C) oil in the steam entering
the heater does not affect
the efficiency of the
heater
(D) it is usually installed
between the feedwater pump
and the boiler.

58. Which one of the following
statements is true?

(A) In the closed heater the
feedwater can be heated
to nearly 212°F. with L.P.
steam.
(B) Scale and oil do not affect the surrender of heat
in the closed heater.
(C) The condensed steam is returned to the boiler
with the feedwater in the
open type heater.
(D) An oil separator is not
necessary in an open feedwater heater that takes
steam from a reciprocating
steam engine.

59. A direct connected steam turbine is NOT applicable

(A) to driving a centrifugal
pump that discharges
against a high head
(B) where a large starting
torque is involved
(C) to driving a large electric generator
(D) to driving a multistage
centrifugal air compressor.

60. In the average non condensing
single-stage impulse type
Terry turbine it is common
practice to

(A) use balance pistons
(B) operate at 2400 rpm
(C) return condensate to the
boiler first passing steam
through an oversize oil
separator

(D) provide means for controlling end thrust of rotor.

61. An intercooler is usually used

(A) in an ammonia compressor discharge line
(B) between stages of a reaction steam turbine
(C) in a two-stage air compressor
(D) with an air compressor only if cylinder jackets are not water cooled.

62. If the cu. ft. of air delivered by a forced draft fan is doubled which one of the following is true:

(A) This necessitates increasing the fan speed to twice its former speed.
(B) The horsepower input also is doubled.
(C) The horsepower input does not change.
(D) The static pressure increases as the cube of the speed.

63. The important factor to be considered regarding an economizer in a large steam generating plant is that

(A) the flue gas temperature can be lowered below the dew point without ill effect
(B) impure feedwater may be used within the tubes without affecting efficiency
(C) the higher the flue gas temperature the smaller will be the thermal saving
(D) the higher the flue gas temperature the greater will be the thermal saving.

64. In a "balanced draft" system the chimney and fan draft are controlled so that the pressure in the combustion chamber is

most nearly

(A) plus 0.1 in. of water
(B) minus 1.0 in. of water
(C) atmospheric
(D) plus 0.1 in. of water.

65. In a steam generating plant a Ringleman chart is very helpful in determining the

(A) alkalinity of boiler water
(B) rate of boiler water feed
(C) percent of CO_2 in the flue gas
(D) visible smoke emitted from a stack.

66. The synchronous speed in rpm of a 4 pole, 3 phase, 60 cycle induction motor running at no load is most nearly

(A) 3600   (C) 1200
(B) 1800   (D) 720.

67. Squirrel cage induction motors are easily identified because they have

(A) two slip rings
(B) commutators
(C) three pairs of brushes
(D) no moving contacts of any kind.

68. A wound rotor induction motor

(A) has copper conductors in the rotor which are not insulated
(B) does not have any slip rings
(C) when running at full speed operates similar to the squirrel cage motor
(D) is not usually started by the use of resistances in the rotor circuit.

69. If an electrical power system uses a balancer set and the system is balanced, then

(A) both machines of balances set operate as generators
(B) both machines of the

balancer set operate as
motors
(C) one machine of the balancer set operates as a
motor driving, the other
as a generator
(D) neither machine will run.

70. D.C. generators that do not
require equalizing connections
when connected in parallel are

(A) series generators
(B) compound generators
(C) shunt generators
(D) none of the above.

71. Assume two D.C. generators are
connected in parallel, and you
want to shut down generator
No. 2. This can be done by

(A) opening the main switch
regardless of ammeter
reading
(B) first tripping circuit
breaker, then opening
main switch
(C) adjusting field rheostat
to strengthen shunt-field
current before opening
main switch
(D) first weakening the shunt-
field current, then open-
ing main switch when am-
meter reads zero.

72. The cost per hour to run a 20
hp. motor having an efficiency
of 80% with electricity at 4¢
per kilowatt hour is approximately (1 hp. = 746 watts)

(A) $.75   (C) $1.00
(B) $.48   (D) $1.50.

73. When a rotary converter is run
from the A.C. side the D.C.
output voltage is

(A) greater than the A.C.
voltage
(B) equal to the A.C.
voltage
(C) smaller than the A.C.
voltage
(D) .707 times the A.C. voltage.

74. A rotary converter operates
from the A.C. side as a synchronous motor at unity power
factor. It is desired to run
it at a leading power factor.
To do this you would

(A) decrease the D.C. field
strength of the machine
(B) increase the D.C. field
strength of the machine
(C) decrease the speed of the
machine
(D) increase the speed of the
machine.

75. If the power factor of a synchronous motor is changed from
lagging to leading, the A.C.
current delivered in the motor

(A) drops to zero and then increases
(B) is increased immediately
(C) increases to a maximum and
then decreases
(D) decreases to a minimum and
then increases.

76. In a triple expansion steam
engine there is

(A) a decrease in wear and
tear of moving parts
(B) a decrease in radiation
losses
(C) increased economy in steam
consumption
(D) increased clearance and
leakage losses.

77. In a uniflow steam engine the
steam is exhausted from the

(A) crank end of the cylinder
(B) center of the cylinder, at
the furthest point from
the heads
(C) head end of the cylinder
(D) third point of the cylinder, measured from the head
end.

78. A pyrometer is used to measure

(A) temperature
(B) steam pressure

(C) draft
(D) quantity of steam in a
 boiler.

79. A Bourdon pressure gauge is
 usually graduated in

 (A) cu. ft. per min.
 (B) lbs. per sq. ft.
 (C) cu. inches per sec
 (D) lbs. per sq. in.

80. The Orsat apparatus is used to

 (A) record stack temperatures
 (B) analyze flue gases
 (C) determine the quality of
 steam
 (D) determine the heat evalu-
 ation of fuel.

Answer Key

1.	D	17.	A	33.	A	49.	A	65.	D
2.	B	18.	C	34.	D	50.	C	66.	B
3.	D	19.	C	35.	B	51.	B	67.	D
4.	A	20.	C	36.	C	52.	D	68.	C
5.	C	21.	B	37.	A and B	53.	C	69.	B
6.	A	22.	B	38.	B	54.	B	70.	C
7.	A and B	23.	B	39.	B	55.	B	71.	D
8.	C	24.	A	40.	C	56.	A	72.	A
9.	D	25.	B	41.	C	57.	D	73.	A
10.	B	26.	D	42.	C	58.	C	74.	B
11.	D	27.	C	43.	B	59.	B	75.	D
12.	D	28.	B	44.	A	60.	B	76.	C
13.	D	29.	A	45.	B	61.	C	77.	B
14.	C	30.	D	46.	B	62.	A	78.	A
15.	B	31.	C	47.	C	63.	D	79.	D
16.	B	32.	A	48.	D	64.	C	80.	B

SAMPLE PRACTICE EXAMINATION 5

HIGH PRESSURE PLANT TENDER

DIRECTIONS: Each question has four suggested answers, lettered A, B, C, and D. Decide which one is the best answer, locate the question number on the sample answer sheet, and with a soft pencil darken the area that corresponds to the answer you have selected. All sample answer sheets follow page 22.

The time allowed for the entire examination is 3½ hours.

1. An H.R.T. boiler is one which is best briefly described as follows:

   (A) horizontal, stationary, internally fired, fire tube
   (B) horizontal, stationary, externally fired, water tube
   (C) horizontal, stationary, externally fired, fire tube
   (D) horizontal, rotating, externally fired, fire tube.

2. A water tube boiler is one which has

   (A) water on the outside of tubes, hot gases on the inside
   (B) scale on the outside of tubes, soot on the inside
   (C) water on the inside of tubes, hot gases on the outside
   (D) none of the above.

3. A Sterling boiler is a

   (A) fire tube boiler with straight tubes
   (B) water tube boiler with straight tubes
   (C) boiler with box headers
   (D) water tube boiler with bent tubes.

4. The boiler drum of a high pressure boiler is usually constructed with

   (A) lap joints
   (B) butt joints
   (C) spot welded joints
   (D) thermit welded joints.

5. In a Sterling boiler, a radiant type superheater is usually located in the

   (A) second pass
   (B) third pass
   (C) first pass
   (D) uptake.

6. The holes in the fire door of a hand fired furnace

(A) are provided to prevent the door from warping due to the heat of the furnace
(B) allow secondary air to enter the furnace
(C) should be kept closed
(D) should be opened wide if the boiler setting is weak.

7. A feedwater pump supplies water at 150 Btu/lb. to a boiler rated at 10,000 boiler horsepower and generating steam at 1250 Btu/lb. Assume that a boiler horsepower is equivalent to the evaporation of 34.5 lbs. of water per hr. under atmospheric pressure at 212 F., water's latent heat of evaporation is 970 Btu/lb., and a gal. of water weighs 8.3 lbs. Approximately how many gal. per hr. must this pump deliver?

(A) 36,000 gph
(B) 83,000 gph
(C) 304,000 gph
(D) 345,000 gph.

8. Approximately how many gal. per min. of feedwater would have to be evaporated in order to produce one boiler horsepower if steam with a total heat content of 1250 Btu/lb. is being generated and feedwater is supplied at 212°F.? Assume that a boiler horsepower is equivalent to the evaporation of 34.5 lbs. of water per hr. under atmospheric pressure at 212°F., the water's latent heat of evaporation is 970 Btu/lb., and a gal. of water weighs 8.3 lbs.

(A) .06 gpm   (C) 4.1 gpm
(B) 3.8 gpm   (D) 31.3 gpm.

9. As received, the heating value (Btu. per lb.) of a good grade of bituminous coal is most nearly

(A) 12,000   (C) 16,500
(B) 14,000   (D) 10,000.

10. A coal with 30% volatile matter on a moisture-free basis would be classed as

(A) bituminous
(B) lignite
(C) anthracite
(D) high-rank semi-bituminous.

11. Which one of the following pieces of equipment can be used for feeding water to an operating boiler?

(A) an accumulator
(B) an underfeed single retort stoker
(C) an ejector
(D) an injector.

12. A check valve is installed on a boiler feedwater line in order to

(A) prevent back-up of blow-down from another boiler
(B) prevent boiler water from returning to the pumping units
(C) prevent the steam in the boiler from mixing with the feedwater
(D) permit repairing gauge glass with the boiler steaming.

13. In a steam duplex boiler feed pump the steam piston diameter is

(A) larger than the water piston diameter
(B) smaller than the water piston diameter
(C) usually equal to the water piston diameter to enable it to feed the boiler
(D) smaller than the water piston diameter because it is double acting.

14. If two boilers were feeding the same steam main and each

boiler gauge showed a pressure of 50 lbs. per sq. in., the steam main pressure would be approximately

(A) 100 lbs. per sq. in.
(B) 52 lbs. per sq. in.
(C) 104 lbs. per sq. in.
(D) 50 lbs. per sq. in.

15. When connecting a Bourdon steam gauge to a boiler, a goose neck or syphon is used to

(A) correct for air in the line
(B) prevent water from getting into the tube
(C) prevent steam from getting into the tube
(D) bring the gauge to correct reading level.

16. A balanced draft device is used to

(A) keep the pressure over the fire at approximately atmospheric pressure
(B) equalize the draft in the boiler passes
(C) keep the pressure under and over the fire the same
(D) balance the pressure between the boiler room and the bottom of the stack.

17. "Draft in inches of water" when referred to the space directly above the fire means

(A) gauge pressure above atmospheric pressure
(B) the difference in pressure between the boiler room and the space to which the draft gauge is connected
(C) the pressure above atmospheric pressure
(D) the difference in pressure between atmospheric pressure and barometic pressure.

18. On high pressure boilers the water column is used in order to

(A) show directly to the operator the level of the water in the tubes
(B) easily find the pressure within the boiler
(C) dampen the oscillation of the water in the gauge glass
(D) prevent the boiler from exploding.

19. Steam at a pressure of 100 lbs. per sq. in. gauge and a temperature of 400 F. is

(A) superheated steam
(B) condensate steam
(C) saturated steam
(D) wet steam.

20. Priming of a boiler is <u>not</u> desirable because

(A) too much make-up water is needed
(B) it will wet the safety valve
(C) it causes boiler water level to rise
(D) steam with the greatest heat content is usually desired.

21. When the water level is no longer visible in a gauge glass it is first best to

(A) start the standby feedwater pump
(B) close the inside stop on the boiler steam line
(C) feed water into the boiler through the city water line
(D) cover the fire with green coal leaving fire doors open and then check water level.

22. The CO_2 in the flue gas is 13% over the fire, 11% in the second pass and 8% at the damper. This most likely indicates

(A) leaks through side walls
(B) not enough air under the fire
(C) too much air under the fire
(D) incomplete combustion.

23. An analysis of flue gas from a hand-fired boiler shows some CO. This indicates

(A) feedwater is too cold
(B) incomplete combustion
(C) good combustion
(D) the furnace temperature is too high.

24. When burning anthracite coal good combustion efficiency is usually associated with

(A) 16% CO_2 and 2% CO
(B) 10% CO_2 and 1% CO
(C) 10% CO_2 and 2% CO
(D) 14% CO_2 and 0% CO.

25. Chemicals most commonly used to soften boiler feedwater are

(A) barium and phosphate
(B) bicarbonate of soda and alum
(C) soda ash and lime
(D) sodium chloride and potash.

26. When a boiler is said to prime it usually means that

(A) it is carrying water over with the steam
(B) steam is just beginning to form after starting up
(C) the boiler is making a drumming noise
(D) the boiler is being slightly overfired.

27. An automatic feedwater regulator is used

(A) to take the place of the boiler feed pump
(B) when the lower water alarm sounds

(C) to maintain a constant water level
(D) to control the chemical treatment of the feedwater.

28. A boiler must be blown down

(A) not more often than once a day
(B) as often as necessary to prevent priming and/or foaming
(C) not more often than once a week
(D) as often as necessary to prevent slagging.

29. The surface blow-off apparatus is commonly used for the purpose of

(A) giving correct water level
(B) removing impurities from the water surface
(C) desuperheating the steam
(D) giving superheated steam.

30. An inclined water tube boiler is equipped with two blowdown lines from a single mud drum. If only one of these blowdown lines is consistently used, this would probably result in

(A) the unused blowdown valves freezing in the closed position
(B) poor water circulation on one side of the boiler
(C) no abnormal condition
(D) an abnormal condition not mentioned above.

31. A steam lance is a device most commonly used for

(A) increasing the draft in the stack
(B) softening scale in the boiler drum
(C) removing soot from the inside of boiler tubes
(D) removing soot from the outside of boiler tubes.

32. In an angle valve, plug valve, boiler blow-off valve

combination the plug valve is usually

(A) mounted next to the boiler, opened last and closed first, in winter time only
(B) mounted furthest from the boiler, opened last and closed first
(C) mounted next to the boiler, opened first and closed last
(D) mounted furthest from the boiler, opened first and closed last.

33. A pop type safety valve is commonly a

(A) spring loaded valve
(B) ball and lever valve
(C) dead weight valve
(D) member with a rupture section.

34. During the hydrostatic test of a boiler each safety valve should

(A) be set at maximum allowable working pressure
(B) be set at 150% of maximum allowable working pressure
(C) be held to its seat by screwing down the compression screw upon its spring
(D) have its valve disc held to its seat by means of a testing clamp.

35. When preparing a boiler for inspection one should proceed as follows:

(A) Take the boiler off the line and empty.
(B) Take the boiler off the line, empty boiler after it cools, clean grates and ash pit and clean the water side and fire side.
(C) Take the boiler off the line and wait for the

inspector.
(D) Wait for the inspector and then do as he directs.

36. Staybolts on an installed boiler are tested on inspection by

(A) tapping on one end of each bolt with a hammer
(B) a nick break test
(C) removing one chosen at random
(D) drilling a longitudinal hole through one chosen at random.

37. Tubes in a water tube boiler are secured into the tube sheets and headers by

(A) locknuts and bushings
(B) welding the ends
(C) expanding the ends with a tube expander
(D) placing a collar over the tube end and peening

38. In an inclined tube water tube boiler circulation in the tubes is caused

(A) by steam formed in the tubes
(B) by use of a pump
(C) by use of an injector
(D) by use of baffles.

39. The temperature of flue gas leaving a straight water tube boiler increases suddenly, and remains high even though there was no increase in firing rate. This most likely indicates failure of the

(A) damper   (C) fan
(B) stoker   (D) baffles.

40. The difference between an open and closed feedwater heater is that

(A) the closed feedwater heater deaerates the water
(B) the closed feedwater is covered and the open heater

is not covered
(C) the open feedwater heater has direct mixing of the steam and water, the closed heater does not
(D) the steam and water are at the same pressure in the closed heater and at different pressures in the open heater.

41. The most efficient method of controlling the volume of air delivered by a fan is by

(A) control of fan outlet damper
(B) varying the fan speed
(C) control of fan intake damper
(D) by-passing the fan.

42. In a steam plant the flue gases usually pass through the (1) induced draft fan, (2) economizer, (3) forced draft fan, (4) boiler, (5) air heater, in the following order:

(A) 3 - 4 - 5 - 1 - 2
(B) 4 - 2 - 5 - 1
(C) 3 - 2 - 1 - 4 - 5
(D) 1 - 5 - 4 - 2 - 3.

43. A chain grate stoker is usually

(A) suitable only for coals with high ash-fusion temperatures
(B) suitable only for small size anthracite with low ash content
(C) for hand operation in small bodies
(D) in need of equal quantity of air introduced into each section of the furnace.

44. If a piece of tramp iron finds its way into the worm of an underfeed stoker, the

(A) alarm would sound but the stoker would continue to operate
(B) shear pin would fall and protect the stoker

(C) combustion efficiency would be decreased
(D) iron would burn when it reached the hottest part of the fire.

45. Draft will vary from the highest positive pressure to the lowest negative pressure in a boiler in the following order:

(A) at the top of the stack, at the bottom of the stack, over fire, under-fire
(B) overfire, underfire, in the second pass, in the first pass, no draft at base of stack
(C) at the breeching, in the second pass, in the first pass, overfire, underfire
(D) underfire, overfire, in the first pass, in the second pass, at the breeching.

46. When a boiler low water alarm sounds you should

(A) cover the fire with green coal
(B) dump the fire
(C) add chemical to the feedwater
(D) start the feedwater pump or increase the boiler water feed.

47. A boiler horsepower is by definition, the evaporator per hour, from and at 212°F. of

(A) 14.7 lbs. of water
(B) 34.5 lbs. of water
(C) 7.5 lbs. of water
(D) 8.33 lbs. of water.

48. A 100 horsepower boiler operating at 200% rating generates the following pounds of steam per hour from and at 212°F.:

(A) 3450   (C) 6900
(B) 9600   (D) 5340.

49. The following grades of fuel oil are commonly used in power boilers:

    (A) nos. 1 and 6
    (B) nos. 5 and 6
    (C) nos. 1 and 2
    (D) nos. 3 and 4.

50. No. 6 or bunker C fuel oil used in a mechanical pressure atomizing burner should preferably be heated to most nearly

    (A) 100°F.   (C) 275°F.
    (B) 200°F.   (D) 300°F.

51. The main reason for recirculating heavy fuel oil before starting a fuel oil burner is to

    (A) allow the burner to come up to speed
    (B) build up pressure of the oil
    (C) build up temperature of the oil
    (D) thoroughly strain the oil before burning.

52. The heating of heavy fuel oil in a storage tank with steam coils is permissible provided the steam pressure does not exceed

    (A) 5 lbs. per sq. in.
    (B) 15 lbs. per sq. in.
    (C) 10 lbs. per sq. in.
    (D) 2 lbs. per sq. in.

53. A fuel oil heating system is made up of an electric preheater and a "below the water line" heater. When starting the oil burners with cold boilers the probability is that

    (A) both heaters will heat the oil before the burners start
    (B) the "below the water line heater" will heat the oil first and then the electric preheater

    (C) the "below the water line heater" will heat the oil before the burners start
    (D) the electric preheater will heat the oil before the burners start.

54. The method of determining when a fuel oil suction strainer should be cleaned is to clean it as soon as

    (A) the oil leaving the strainer appears dirty
    (B) sparklers appear in the flame
    (C) the pressure drop across the strainer becomes excessive
    (D) the flame gets smoky.

55. The primary purpose of a barometric damper is to

    (A) control the draft pressure in the boiler and furnace
    (B) circulate cellar air through the chimney to cool it
    (C) provide a means of cleaning the flue
    (D) dampen size of the flame.

56. When taking a CO_2 reading of an operating oil fired boiler

    (A) the burner should be overfired
    (B) the burner should have just been placed in operation
    (C) the sample should be taken from the flue between the barometric damper and the base of the chimney
    (D) the sample should be taken at the breeching of the boiler before it mixes with any other gases.

57. In a No. 6 fuel oil burner installation high and low suction lines are commonly installed in storage tanks in order to

(A) maintain oil temperature at the burners by recirculation through the low suction

(B) make sure all the oil can be withdrawn from the tank when the supply runs low

(C) maintain an oil supply if the water or heavy sludge gets into the bottom of the tank

(D) provide an oil supply for overfiring the boilers.

58. A checkerboard floor construction in the furnace of a heavy fuel oil burner is intended to

(A) eliminate the need for floor refractory

(B) trap and drain off any unburned fuel oil

(C) properly distribute the secondary combustion air into the combustion chamber

(D) absorb expansion of refractory.

59. If the flame in the combustion chamber of an oil fired boiler (mechanical pressure type burner) smokes badly, the direct trouble is most likely

(A) insufficient combustion air

(B) suction strainer is dirty

(C) flame impingement

(D) oil pressure at the nozzle is only 110 psi.

60. If the optical system of a photo-electric smoke meter and alarm is not kept clean, it will

(A) read too low

(B) read too high

(C) short circuit and do damage to the instrument

(D) read normally without error.

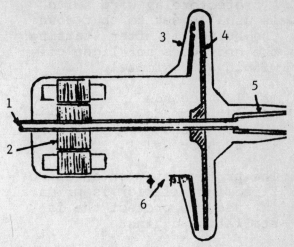

Questions 61-64 refer to the above sketch.

61. The sketch shown represents

(A) a mechanical pressure type oil burner

(B) a rotary cup oil burner

(C) outside mixing air atomizing oil burner

(D) inside mixing steam atomizing oil burner.

62. Atomization is mainly due to the action of

(A) 4   (C) 1
(B) 6   (D) 5.

63. Primary air pressure is produced by the action of

(A) 4   (C) 6
(B) 5   (D) 2.

64. Oil enters the burner at

(A) 6   (C) 4
(B) 5   (D) 1.

65. The magnetic oil valve which is interlocked with the primary air supply should be installed between the

(A) preheater and oil burner

(B) oil burner and furnace

(C) suction heater and preheater

(D) pressure regulator and burner.

66. A protectorelay warp switch
    is usually set to shut down
    the burner if after starting
    the unit does not light off
    within approximately

    (A) 120 seconds
    (B) 145 seconds
    (C) 30 seconds
    (D) 80 seconds.

67. The protectostat on an oil
    fired installation is primarily
    intended to protect the in-
    stallation against

    (A) low CO
    (B) flame failure
    (C) failure of reset to
        operate
    (D) pressurestat failure.

68. On a horizontal rotary cup
    type burner the Vaporstat
    should be electrically inter-
    locked with the

    (A) protectostat

(B) electric oil preheater
(C) oil magnetic valve
(D) low water cutoff.

69. If a deposit of carbon appears
    on the rear wall of the com-
    bustion chamber of a mechani-
    cal pressure type oil burner,
    a good procedure is to

    (A) change the brick in the
        rear wall
    (B) enlarge the combustion
        chamber
    (C) cut down the air supply
    (D) try a wider angle nozzle.

70. Commercial fuel oil standards
    specify a minimum "flash
    point" because oils of lower
    than minimum flash point are
    too

    (A) difficult to atomize
    (B) viscous in cold weather
    (C) dangerous to handle
    (D) difficult to ignite.

Answer Key

1.	C	15.	C	29.	B	43.	A	57.	C
2.	C	16.	A	30.	B	44.	B	58.	C
3.	D	17.	B	31.	D	45.	D	59.	A
4.	B	18.	C	32.	C	46.	D	60.	B and A
5.	C	19.	A	33.	A	47.	B	61.	B
6.	B	20.	D	34.	D	48.	C	62.	D
7.	A	21.	D	35.	B	49.	B	63.	A
8.	A	22.	A	36.	A	50.	B	64.	D
9.	B	23.	B	37.	C	51.	C	65.	A
10.	A	24.	D	38.	A	52.	C	66.	D
11.	D	25.	C	39.	D	53.	D	67.	B
12.	B	26.	A	40.	C	54.	C	68.	C
13.	A	27.	C	41.	C and B	55.	A	69.	D
14.	D	28.	B	42.	B	56.	D	70.	C

SAMPLE PRACTICE
EXAMINATION 6

HIGH PRESSURE PLANT TENDER

DIRECTIONS: Each question has four suggested answers, lettered A, B, C, and D. Decide which one is the best answer, locate the question number on the sample answer sheet, and with a soft pencil darken the area that corresponds to the answer you have selected. All sample answer sheets follow page 22.

The time allowed for the entire examination is 3 hours.

1. Anthracite coal contains

   (A) no hydrocarbons
   (B) no hydrogen
   (C) a high percentage of fixed carbon
   (D) usually high percentage of sulphur.

2. Bituminous coal as received or when burnt

   (A) is considered a hard coal
   (B) has a high percentage of volatile matter
   (C) produces no clinkers
   (D) contains no pitch.

3. Barley coal is also known as

   (A) buckwheat No. 1
   (B) buckwheat No. 2
   (C) buckwheat No. 3
   (D) culm.

4. Coke is usually made from

   (A) bituminous coal
   (B) anthracite coal
   (C) peat
   (D) bagasse.

5. The fuel which contains the greater heating value in Btu's per lb. is

   (A) coal
   (B) natural gas
   (C) tan bark
   (D) No. 6 fuel oil.

6. For the complete and efficient combustion of powdered coal

   (A) low furnace temperatures are essential
   (B) high furnace temperatures are essential
   (C) cold outside air is used
   (D) natural gas is mixed with it.

7. In firing No. 6 fuel oil

   (A) it is atomized as it leaves the storage tank
   (B) it is mixed with No. 2 oil before it is atomized
   (C) it is heated as it comes from the storage tank and then atomized
   (D) it is dehydrated before it is preheated.

8. The protectostat when installed in a fully automatic
rotary cup burner is used for

(A) hi-lo flame control
(B) flame failure detection
(C) preheating the fuel oil
(D) registering the furnace
temperature.

9. The tools required to fire
bituminous coal properly are

(A) shovel, lazy bar, rake,
slice bar and hoe
(B) shovel, hoe and rake
(C) rake, hoe and slice bar
(D) lazy bar, shovel and hoe.

10. The most generally used method
of firing bituminous coal is

(A) the even spread method
(B) carrying a level 3" depth
of fire bed
(C) the coking method
(D) filling in only the dead
spots in the fire bed.

11. Draft is measured in

(A) inches
(B) feet of mercury
(C) inches of mercury
(D) inches of water.

12. The steam gauge is graduated
in

(A) lbs. pressure
(B) in. of pressure
(C) lbs. per sq. in. pressure
(D) lbs. per sq. ft. of
pressure.

13. The injector is operated by

(A) steam
(B) oil
(C) water
(D) natural gas and water.

14. A positive "draft indicator"
reading over the grate in a
boiler operating with forced
draft means that

(A) the draft is sufficient
(B) the draft should be de-
creased
(C) you have a smokeless fire
(D) the damper is opened wide.

15. The term "balanced draft"
means that

(A) the draft pressure in the
wind box is negative
(B) the draft pressure in the
breeching is positive
(C) the draft pressure in the
first pass of the boiler
is positive
(D) the draft pressure in the
combustion chamber is ap-
proximately atmospheric.

16. A Pensky-Martens tester is
used to

(A) find the flash point of
a fuel oil
(B) find the moisture content
of a coal
(C) find the pour point of a
fuel oil
(D) find the nitrogen content
of a fuel oil.

17. The function of a fuel oil
burner is to

(A) feed liquid fuel to the
fire
(B) preheat the fuel oil
(C) atomize the fuel oil to
as nearly a vapor state as
possible
(D) give a blow pipe action of
the flame in combustion
chambers of large capacity.

18. In large hand-fired longitudinal drum water tube boilers
the try-cocks are usually
located

(A) in back of the boiler
(B) on the side of the boiler
(C) on top of the boiler
(D) in front of the boiler.

19. In water tube boilers the hard scale is removed from the inside of the tubes by

    (A) running hot water through them
    (B) turbining, brushing and flushing with water
    (C) using a steam lance
    (D) using an acid solution.

20. An advantage of a Scotch-Marine type boiler is

    (A) that it requires a brick setting
    (B) that it has no radiation losses
    (C) that it has large steaming capacity for the space occupied
    (D) that its circulation is always positive.

21. Soot blowers are used with

    (A) Scotch-Marine boilers
    (B) Babcock and Wilcox boilers
    (C) Kewanee boilers
    (D) Manning boilers.

22. Complete combustion requires that all carbon be converted to CO_2, all sulphur to SO_2, and all hydrogen to H_2O. The percent of excess oxygen that must usually be supplied to the furnace to insure complete combustion is

    (A) 20   (C) 5
    (B) 40   (D) 60.

23. The ignition arch is usually used in a furnace equipped with

    (A) an underfeed stoker
    (B) a multiple retort stoker
    (C) a chain grate stoker
    (D) a sprinkler type stoker.

24. The most desirable combustion rate for low-fusion-ash coals is around

    (A) 10 lbs. per sq. ft. of grate per hr. for continuous ratings
    (B) 35 lbs. per sq. ft. of grate per hr. for continuous ratings
    (C) 65 lbs. per sq. ft. of grate per hr. for continuous ratings
    (D) 70 lbs. per sq. ft. of grate per hr. for continuous ratings.

25. In a steam generating plant the economizer is used to

    (A) superheat the steam
    (B) de-superheat the steam
    (C) heat feed water
    (D) collect fly-ash.

26. The main purpose for chemically testing feedwater before it enters a steam generating boiler is to

    (A) give it a color for test purposes
    (B) make it hard
    (C) prevent rusting the boiler shell
    (D) get rid of hard scale forming materials.

27. The surface blow-off in a steam generating boiler is located

    (A) in the rear of the boiler near the mud drum
    (B) near the surface of the water
    (C) near the first row of tubes
    (D) on top of the boiler drum.

28. In a high pressure steam generating plant which operates at full load for 16 hours out of 24 the boilers are usually blown down during

    (A) the 12 p.m. to 8 a.m. shift
    (B) the 8 a.m. to 4 p.m. shift
    (C) the 4 p.m. to 12 p.m. shift
    (D) any period of operation that the stationary fireman decides.

Questions 29 to 38 are based upon the sketch shown below.

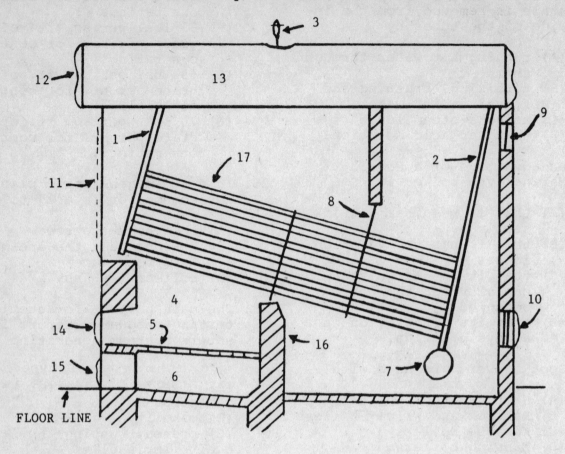

FLOOR LINE

29. The above sketch is that of a

(A) Scotch-Marine boiler
(B) locomotive boiler
(C) Robb-Mumford boiler
(D) Babcock and Wilcox boiler.

30. The fire door is numbered

(A) 14   (C) 11
(B) 15   (D) 10.

31. The furnace is numbered

(A) 6   (C) 4
(B) 1   (D) 8.

32. The mud drum is numbered

(A) 13   (C) 17
(B) 7   (D) 3.

33. The front header is numbered

(A) 11   (C) 1
(B) 8   (D) 2.

34. The baffle is numbered

(A) 17   (C) 16
(B) 8   (D) 12.

35. The damper is numbered

(A) 9   (C) 3
(B) 10   (D) 6.

36. The bridge wall is numbered

(A) 8   (C) 5
(B) 16   (D) 17.

37. The soot door is numbered

(A) 9   (C) 11
(B) 7   (D) 10.

38. The manhole is numbered

(A) 13   (C) 12
(B) 14   (D) 9.

39. The safety valves in high

pressure steam generating oilers should be set by the

(A) watch engineer
(B) chief stationary engineer of the plan
(C) inspector of the Department of Housing and Buildings
(D) inspector of the Department of Water Supply, Gas and Electricity.

40. If a steam driven centrifugal boiler feed pump is in operation but it delivers no water, the trouble generally is that

(A) the impeller is fouled
(B) it is air bound
(C) the shaft is broken
(D) the coupling connecting the pump to the turbine is broken.

41. Small holes are drilled through the ends of screw stays in order to

(A) lighten the stay
(B) save metal
(C) strengthen the stay
(D) indicate by a leak through the holes a break in the stay.

42. Which of the following statements is correct?

(A) The main purpose of a steam jet in connection with "smokeless furnaces" is to mix the air and gases and insure intimate mixture of the products of combustion.
(B) The steam from a steam jet increases the calorific value of the fuel.
(C) The steam from a steam jet is a supporter of combustion.
(D) The most inefficient jets are those based on the injector or siphon principle.

43. In a duplex boiler feed pump

(A) the steam valves have lap
(B) the steam cylinder is larger than the water cylinder
(C) the steam valves have no lost motion
(D) the steam valve is usually on the piston type.

44. The purpose of an air chamber on the discharge side of a duplex boiler feed pump is to

(A) collect the water
(B) increase the water pressure
(C) mix air and water
(D) insure a uniform flow of water.

45. In order to overcome flame impingement on the bridge wall during the operation of a mechanical pressure type burner the usual procedure is to

(A) increase the size of tip
(B) reset the distance the barrel of the burner extends into the furnace
(C) reset the diffuser plate
(D) increase the oil pressure.

46. 34.5 lbs. of water evaporated per hr. from and at 212°F. is by definition

(A) a boiler horsepower
(B) the factor of safety of a boiler
(C) the factor of evaporation
(D) the ft. lbs. in a horsepower.

47. When burning bituminous coal, the general practice is to carry a thickness of fire equal to

(A) 2 in.   (C) 6 in.
(B) 14 in.   (D) 10 in.

48. In steam boiler operation the following sizes of anthracite coal are usually known as "steaming sizes":

(A) lump, chestnut and stove
(B) chestnut, pea and stove
(C) pea, buckwheat and culm
(D) buckwheat, lump and chestnut.

49. A Sterling high pressure boiler has

(A) straight tubes with headers
(B) bent tubes
(C) straight tubes with serpentine headers

(D) only two drums.

50. In power plant work coals are usually ranked as follows:

(A) anthracite, semi-anthracite and bituminous
(B) anthracite, semi-bituminous and bituminous
(C) bituminous, anthracite and semi-lignite
(D) anthracite, bituminous, sub-bituminous and lignite.

Answer Key

1.	C	11.	D	21.	B and C	31.	C	41.	D
2.	B	12.	C	22.	A	32.	B	42.	A
3.	C	13.	A	23.	C	33.	C	43.	B
4.	A	14.	B	24.	B	34.	B	44.	D
5.	D	15.	D	25.	C	35.	A	45.	B
6.	B	16.	A	26.	D	36.	B	46.	A
7.	C	17.	C	27.	B	37.	D	47.	D
8.	B	18.	D	28.	A	38.	C	48.	C
9.	A	19.	B	29.	D	39.	C	49.	B
10.	C	20.	C	30.	A	40.	B	50.	D

SAMPLE PRACTICE EXAMINATION 7

HIGH PRESSURE PLANT TENDER

DIRECTIONS: Each question has four suggested answers, lettered A, B, C, and D. Decide which one is the best answer, locate the question number on the sample answer sheet, and with a soft pencil darken the area that corresponds to the answer you have selected. All sample answer sheets follow page 22.

The time allowed for the entire examination is 3 hours.

1. If three boilers were feeding the same steam main, and each boiler gauge showed a pressure of 50 lbs. per sq. in., a gauge connected to the main would read approximately

   (A) 150 lbs. per sq. in.
   (B) 53 lbs. per sq. in.
   (C) 50 lbs. per sq. in.
   (D) 156 lbs. per sq. in.

2. A type of coal occupies 40 cu. ft. per ton. The capacity of a bunker that is 16 ft. high, 60 ft. long and 20 ft. wide when filled level would be

   (A) 630   (C) 840
   (B) 360   (D) 480.

3. The expression "draft in inches of water" when referred to the space directly above the fire bed means

   (A) pressure above atmospheric pressure
   (B) gauge pressure above atmospheric pressure

   (C) difference in pressure between the boiler room and the space to which the draft gauge is connected
   (D) difference between the barometer pressure and atmospheric pressure.

4. The usual elliptical-shaped inside hand hole cover is only held in place on a steaming boiler

   (A) by a bolt
   (B) by a gasket
   (C) by a bolt and yoke
   (D) by a bolt nut yoke and steam pressure.

5. Minimum excess air is required with which of the following method of firing?

   (A) hand
   (B) pulverized coal
   (C) underfeed stoker
   (D) spreader stoker.

6. If the diameters of a stay rod is doubled its staying force is

91

(A) unchanged
(B) increased two times
(C) increased four times
(D) decreased two times.

7. A three pass straight tube water tube steam boiler is properly equipped with automatic steam type soot blowers. When blowing the tubes the operator should

(A) first blow soot from tubes in first pass
(B) first blow soot from tubes in second pass
(C) first blow soot from tubes in third pass after opening the up-take damper fully
(D) first blow soot from tubes in the first pass after opening the up-take damper fully.

8. Safety valves are required on the superheater discharge header mainly to

(A) prevent burned out tubes
(B) provide boiler safety valve capacity
(C) be easily accessible
(D) none of the above.

9. An analysis of flue gas from a hand-fired boiler shows some CO. This means

(A) incomplete combustion
(B) good combustion
(C) feedwater is too cold
(D) the furnace temperature is too high.

10. A pop safety valve is commonly a

(A) member with a ruptured section
(B) dead weight valve
(C) ball and lever valve
(D) spring loaded valve.

11. A boiler has a heating surface of 1000 sq. ft. Its rating in boiler horsepower is approximately

(A) 1000   (C) 100
(B) 500   (D) 200.

12. One advantage of steam-atomization of fuel oil is that

(A) the oil burns better
(B) comparatively dirty oil can be used
(C) they are quiet in operation
(D) the quantity of steam used is about .1% of the boiler output.

13. When the orifice of a mechanical pressure type oil burning system is changed and the oil pressure is kept constant

(A) the degree of atomization of the oil is greatly affected
(B) the quantity of oil flow varies
(C) the oil flow is tripled if the orifice area is doubled
(D) the fuel oil temperature must be changed.

14. One important advantage of a poppet valve over a Corliss type valve in steam engines is

(A) they are easier to build
(B) the steam chest construction is simpler
(C) that superheater steam may be used with them
(D) that a governor is not required to control speed.

15. In a bleeder turbine used for process steam and power, as the amount of steam used for process is increased

(A) the electrical load is increased
(B) the the electrical load is decreased
(C) the electrical load need not necessarily change
(D) the process steam pressure increases.

16. One horsepower acting for one hour is equal to the following number of Btu:

    (A) 2769   (C) 2545
    (B) 5000   (D) 2000.

17. The auxiliary exhaust valves on a non-condensing uniflow engine eliminates

    (A) the need for exhaust ports
    (B) excessive compression pressure
    (C) wire drawing at admission
    (D) the possibility of using the engine for condensing operation.

18. A low speed fan is desirable for induced draft due to

    (A) the presence of cinders and fly ash in the flue gas
    (B) its high tip speed
    (C) its high static efficiency
    (D) the fact that the gases are slowed down after going through the boiler.

19. The difference between an open and closed feedwater heater is that

    (A) the open feedwater heater is not covered and the closed feedwater heater is covered
    (B) the open heater has direct mixing of the steam and water, the closed heater does not
    (C) the closed feedwater heater deaerates the water
    (D) the steam and water are at different pressures in the open heater, at the same pressure in the closed heater.

20. If the mean effective pressure of a constant speed engine be doubled, the indicated horsepower is

    (A) halved
    (B) unchanged

    (C) increased 90%
    (D) doubled.

21. In a steam duplex boiler feed pump the water piston diameter is

    (A) smaller than the steam piston diameter
    (B) larger than the steam piston diameter
    (C) larger than the steam piston diameter so that a higher water pressure can be reached
    (D) usually equal in diameter to enable it to feed the boiler.

22. A check valve is installed on a feedwater line

    (A) to prevent the steam in the boiler from mixing with the feedwater
    (B) to prevent the return flow of the boiler water to the pumping units
    (C) so that the gauge glass may be repaired with the boiler steaming
    (D) to prevent back-up of blowdown from another boiler.

23. The equipment required in the exhaust line from reciprocating steam engines when the exhaust is used for heating is

    (A) an oil separator, a grease extractor and feedwater heater
    (B) only a steam line to the radiators
    (C) a condenser and muffler
    (D) an oil separator, muffler and regulating valves.

24. The advantage of a direct acting steam pump is that

    (A) it is thermally efficient
    (B) higher water pressure can be developed
    (C) it is simple and rugged
    (D) no lubricating oil gets into the water.

25. A foot valve is used in the suction line of a centrifugal pump

    (A) to make sure the water leaves through the discharge line
    (B) to keep the pump primed
    (C) only when oil is being pumped
    (D) so as to start with minimum load.

26. Which of the following tends to keep condenser pressure low?

    (A) deaerating hotwell
    (B) relief valve
    (C) baffles
    (D) steam jet air ejector.

27. An evaporator is used in a large capacity condensing steam power plant to

    (A) provide make-up water
    (B) provide drinking water
    (C) to increase the boiler capacity
    (D) treat water before it is put through a zeolite tank.

28. Steam with a quality of 80% means that for 1 lb. of wet steam there is

    (A) .8 lb. of water
    (B) .2 lb. of water
    (C) .2 lb. of dry steam
    (D) 1.8 lbs. of dry steam.

29. A steam gauge should be connected to a high pressure steam line

    (A) by insulated piping
    (B) by means of a siphon
    (C) directly to an opening at the top of the pipe
    (D) directly to an opening in the side of the pipe.

30. The speed of a forced draft fan is doubled. As a result

    (A) the cu. ft. per min. delivered is unchanged
    (B) the cu. ft. per min. delivered is doubled
    (C) the power input is doubled
    (D) the static head is tripled.

31. A pump requiring 15 kw is driven by a 25 hp. motor. The motor input is most nearly

    (A) 35 kw   (C) 15 kw
    (B) 20 kw   (D) 7.5 kw.

32. When the discharge valve on an operating centrifugal pump is closed

    (A) the power input is zero
    (B) the power input is a maximum
    (C) the power input is a minimum
    (D) the head developed is a minimum.

Questions 33 to 37 are based on the drawing below.

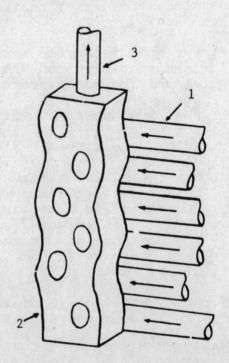

33. The drawing represents one section of a

    (A) water wall tube
    (B) sectional header
    (C) girth superheater
    (D) dry pipe.

34. Tubes (1) are fastened into (2) by

    (A) welding   (C) rolling
    (B) screwing   (D) soldering.

35. Tubes (1) lead from the

    (A) drum
    (B) lower header
    (C) feedwater inlet
    (D) mud drum.

36. Tube (3) connects to the

    (A) superheater
    (B) boiler drum
    (C) safety valve
    (D) dry pipe.

37. Circulation is caused in the direction indicated by the arrows

    (A) by use of a pump
    (B) because steam is formed in the tubes
    (C) because the uprights are curved
    (D) by use of an injector.

38. Two unloaders are used on an air compressor

    (A) to change load stepwise
    (B) to have one as standby
    (C) as a safety feature if one fails to operate
    (D) to make two-stage compression without the use of an inter-cooler.

39. The water column is used on high pressure boilers in order to

    (A) prevent the boiler from exploding
    (B) dampen the oscillation of the water in the gauge glass
    (C) easily find the pressure within the boiler
    (D) give the operator the level of the water in the tubes.

40. The proximate analysis of a fuel reports

    (A) the percentage by weight of moisture, fixed carbon, volatile matter and ash
    (B) the percentage by volume of moisture, carbon, volatile matter and ash
    (C) the percentage of weight of sulphur, iron, moisture, volatile matter and ash
    (D) the percentage by volume of moisture, iron, ash and volatile matter.

41. Smoke from a coal-burning boiler can be diminished by

    (A) increasing the steam pressure
    (B) introducing air above the fire
    (C) decreasing the draft
    (D) decreasing the travel of the gases through the boiler.

42. The "steam loop" is used

    (A) for automatically returning high-pressure drips to the boiler
    (B) to return the drips to a trap
    (C) to supply steam to a separator
    (D) to prevent low-pressure drips from returning to the steam engine.

43. A "telescope oiler" is usually used

    (A) to lubricate bearings of large steam turbines
    (B) when the lubricant is grease

(C) on boiler feed pumps
(D) to lubricate a crosshead and guides.

44. Three solid lubricants which are usually mixed with grease or oil are

(A) mica, powdered granite and shale
(B) soapstone, dry graphite and mica
(C) shale, soapstone and dry graphite
(D) limestone, slate and schist.

45. In order to increase the economy of piston engines you should

(A) increase the back pressure
(B) not reheat receivers
(C) increase the rotative speed
(D) desuperheat the steam supplied to the engines.

46. In order to protect the wrought-steel superheater in a boiler that is being warmed before steam is generated, the operator usually

(A) directs a blast of cold air on the superheater
(B) diverts the direction of the heated gases
(C) disconnects the superheater and replaces when boiler is steaming
(D) does nothing.

47. If the indicator on a steam gauge mounted on the front of a boiler pointed to the number 76, it should be read as

(A) 76 lbs. pressure
(B) 90 lbs. per sq. in. gauge
(C) 76 lbs. per sq. in. gauge
(D) 76 lbs. per sq. ft. of boiler surface.

48. In fitting brushes to a commutator of a D.C. motor, use should be made of

(A) a half-round file for shaping
(B) a file for shaping and emery cloth for the final fitting
(C) only a running-in period of several hours with light load
(D) sandpaper.

49. If two 30-ampere fuse strips are used side by side in a cartridge fuse, the fuse will then blow approximately at

(A) 60 amperes
(B) 30 amperes
(C) some value more than 60 amperes
(D) 10% greater than the sum of the two.

50. A balancer set in a power generating plant is

(A) used in A.C. power generation
(B) used in two-wire D.C. generation
(C) used in three-wire D.C. generation
(D) used to convert A.C. to D.C.

Answer Key

1.	C	11.	C	21.	A	31.	B	41.	B
2.	D	12.	B	22.	B	32.	C	42.	A
3.	C	13.	B	23.	D	33.	B	43.	D
4.	D	14.	C	24.	C	34.	C	44.	B
5.	B	15.	C	25.	B	35.	B	45.	C
6.	C	16.	C	26.	D	36.	B	46.	B
7.	D	17.	B	27.	A	37.	B	47.	C
8.	A	18.	A	28.	B	38.	A	48.	D
9.	A	19.	B	29.	B	39.	B	49.	A
10.	D	20.	D	30.	B	40.	A	50.	C

SAMPLE PRACTICE
EXAMINATION 8

HIGH PRESSURE PLANT TENDER

DIRECTIONS: Each question has four suggested answers, lettered A, B, C, and D. Decide which one is the best answer, locate the question number on the sample answer sheet, and with a soft pencil darken the area that corresponds to the answer you have selected. All sample answer sheets follow page 22.

The time allowed for the entire examination is 3½ hours.

1. When a body of cold air is heated its

   (A) relative humidity is increased
   (B) wet bulb temperature is decreased
   (C) specific volume is decreased
   (D) dew point is unchanged.

2. In the usual classification distinguishing one type of coal from another, soft coal (bituminous) compared to hard coal (anthracite) always has a

   (A) lower heat content
   (B) higher percentage of volatile
   (C) higher ratio of carbon to hydrogen
   (D) higher melting point of ash.

3. Vibrating reed tachometers are most frequently used on

   (A) power factor regulators
   (B) steam turbines

   (C) battery charging equipment
   (D) voltage regulators.

4. When a boiler is said to be prime it usually means that

   (A) it is carrying water over with the steam
   (B) the water in the gauge glass is unstable due to erratic circulation
   (C) the boiler is making a drumming noise
   (D) there is an intermittent puffing of the oil fire.

5. In a modern electric elevator at its highest speed, the emergency over-speed governor usually acts

   (A) electrically, to throw the brakes on the elevator motor
   (B) electrically, to stop the car by throwing motor current off
   (C) mechanically, to seize the lifting cables to stop the car

(D) through governor cable, to increase friction on guide rails.

6. The temperature in a steam turbine condenser is 80° F., and the vacuum referred to 30" barometer is 24". This indicates

(A) insufficient turbine exhaust nozzle area
(B) too much air in condenser
(C) inefficient condensate pump performance
(D) condenser too small for requirements.

7. A Van Stone flange

(A) has a square tongue and groove, and is screwed on the pipe
(B) is a loose flange with 1/8" vee opposite grooves
(C) is a loose steel flange, the halves of which do not touch
(D) is a cast iron flange for low pressure piping.

8. If a sulphur candle is burned in an atmosphere containing ammonia vapor as a test for presence of ammonia

(A) the fumes will be white
(B) the fumes will be brown
(C) the odor will be sharp and acid
(D) the odor will be sweetish.

9. "Degree days" as commonly used is

(A) the ratio of actual fuel used to the theoretical amount required by the weather
(B) the sum of the differences between 65 F. and the average lower outdoor temperature for each day
(C) the number of days times the difference between the actual temperature and 70°F.
(D) the accumulated differences between the actual average

temperature and 0°F.

10. In a two-cycle diesel engine

(A) the exhaust gases are hotter than in a four-cycle engine
(B) valves must be placed in the piston head to admit air to the cylinder
(C) either a hot bulb or glow plugs for starting must be provided
(D) fresh air is taken in and the burned gases are exhausted at the same time.

11. A Hartford loop is a

(A) method of arranging piping to permit expansion
(B) method of arranging boiler piping to equalize boiler pressure
(C) water seal to prevent entry of air into a heating system
(D) safety method of piping returns to a boiler.

12. The piece of equipment known as a dry pipe is most often used with or in a

(A) boiler   (C) air compressor
(B) turbine   (D) vacuum pump.

13. In the usual furnace of a stoker fired boiler equipped with balanced draft and forced draft in good repair, the draft in the furnace should be maintained in inches of water at

(A) .04   (C) 1.0
(B) .003   (D) .4.

14. You would be most likely to find an unloader on

(A) a switch board
(B) an air compressor
(C) a vacuum pump
(D) a centrifugal pump.

15. Spalling is likely to result from

(A) too rapid and too extreme changes in furnace temperature

(B) too heavy a load on the ventilation fan

(C) defective grease trap

(D) leaky packing at shaft.

16. The vent from plumbing fixtures of a multi-story building is

(A) taken from the bottom of the fixture trap into the vent stack

(B) taken from the top of the fixture trap into the soil lime

(C) taken to the roof from the soil

(D) taken into the vent stack from a point between the fixture trap and the soil stack.

17. If the vacuum gauge on an electric driven vacuum pump of a two-pipe vacuum heating system shows a vacuum lower than is usual or desired, you should first

(A) take the pump apart for examination

(B) use a cold water spray in the suction line

(C) check the temperature of the returns

(D) hunt for leaking traps.

18. For the same diameter of piston the connecting rod bearing is usually largest in

(A) an ammonia compressor

(B) a single stage air compressor

(C) a gasoline engine

(D) a diesel engine.

19. The maximum described amount of carbon dioxide that a flue-gas analyzer should be expected to indicate, for a boiler fired with No. 6 fuel oil, is most nearly

(A) 6%   (C) 20%
(B) 13%   (D) 23%.

20. Firebrick should always be laid

(A) in header courses

(B) with not less than 1/8 inch mortar to allow expansion

(C) with only the cement that will stick to the brick

(D) as facing of first, second, and third passes of boiler.

21. A Bourdon tube is

(A) the tube inserted in a duct to measure velocity and pressure

(B) a curved oval tube which tends to straighten when the inside pressure is increased

(C) a U tube partly filled with mercury or water and used to measure light pressures

(D) the middle part of a venturi, in-the-line, flow meter.

22. The clearance in a uniflow engine is larger than in a counterflow engine because

(A) the release pressure is less

(B) it is the practice to use poppet valves

(C) compression begins earlier

(D) the exhaust eccentric is set back.

23. The chemical Freon, or dichlorodifluoromethan, is commonly used

(A) in flue gas analysis

(B) in refrigeration

(C) in treating boiler feed water

(D) as a constituent of commercial gas.

24. In a straight tube water tube boiler that is vertically baffled with three passes a convection type superheater is usually placed

    (A) between first and second row of tubes
    (B) on the furnace side walls
    (C) between first and second pass
    (D) between second and third pass.

25. In the usual type of B & W straight tube boiler installation operating on natural draft, with its setting and horizontal breeching reasonably tight, when the plant is operating at about 50% of maximum capacity, you would expect to find the greatest draft, as measured by draft gauge

    (A) in the ash pit
    (B) in the furnace
    (C) in the last pass
    (D) at the base of the chimney.

26. A P trap for plumbing fixtures, as distinct from an S trap, is

    (A) a syphon type of trap
    (B) a jet type of trap
    (C) a trap discharging downward
    (D) a trap discharging horizontally.

27. The statement that is true of CO_2 refrigeration, as compared to ammonia refrigeration, is

    (A) the raw refrigerant costs less, and less power is used per ton
    (B) the expansion temperature is lower for the same pressure in the cooler, and less power is used per ton
    (C) less pressure is required, and the refrigerant is less toxic.

    (D) the refrigerant is less toxic, and more power is used per ton.

28. Of the following types of manual valves the one it would be best to use for applications requiring throttling of fluid flowing to a piece of equipment is the

    (A) ball valve
    (B) check valve
    (C) gate valve
    (D) globe valve.

29. A Dutch oven is

    (A) used to test the Btu value of fuel
    (B) the top of brickwork in the rear of an h.r.t. boiler
    (C) composed of a double baffle bridge wall.
    (D) an extended furnace with refractory cover.

30. A pitot tube is most commonly used in

    (A) ventilating work
    (B) plumbing
    (C) steam heating systems
    (D) flue gas analysis.

31. The oil for cylinder lubrication of a steam engine using steam at 100 lbs. pressure

    (A) should preferably contain tetraethyl lead
    (B) must always be straight mineral oil
    (C) is often compounded with fatty oil
    (D) must be asphaltic base.

32. In air conditioning for offices the relative humidity is usually maintained at

    (A) 10%   (C) 50%
    (B) 20%   (D) 75%.

33. The theoretical water hp. of a pump is found as follows:

(A) ft. of head times lbs. of water per sec. divided by 30
(B) ft. of head times lbs. of water per min. divided by 33,000
(C) lbs. head times velocity in ft. per min. divided by 550
(D) ft. of head times lbs. of water per min. divided by 550.

34. A Monolithic boiler baffle

(A) is built up of tile shapes
(B) is L shaped to gain high radiant heat transfer
(C) is limited to horizontal baffles
(D) is formed in one piece.

35. The heat of the liquid as shown in steam tables is

(A) 212°F. less than 32°F.
(B) the total heat less the latent heat
(C) the total heat plus 32°F.
(D) the latent heat less 32°F.

36. The uniflow is more economical than a counterflow engine because

(A) there is less initial condensation
(B) the cylinder is smaller
(C) the release pressure is lower
(D) clearance is less.

37. The usual type of thermostatic radiator trap as applied to a two-pipe vacuum heating system

(A) is affected by and operates according to radiator pressure only
(B) depends for its operation upon both pressure and temperature of steam or water in the radiator
(C) is designed to permit the escape of air only
(D) acts to maintain constant room temperature.

38. The loss due to moisture in flue gas as a result of hydrogen in the fuel is greatest in

(A) No. 6 fuel oil
(B) anthracite coal
(C) bituminous coal
(D) Pocahontas coal.

39. Excess air in a Heine boiler installation with flat grates is usually restricted so that

(A) air leaking into the setting above the furnace
(B) air required to combine with the CO formed in the fuel bed
(C) air admitted over the grated through the fire door
(D) air over and above the amount required for complete combustion.

40. A shipper wheel is a

(A) safety wheel inside car to apply safety brakes
(B) drum on which is wound safety cable
(C) wheel in car, connected to shaft to control contacts governing elevator movements
(D) wheel on elevator machine which governs the direction of car movement.

41. A pressure atomizing oil burner produces a

(A) hollow cone of atomized oil
(B) solid cone of atomized oil
(C) spiral
(D) cylindrical flame.

42. The throw of the eccentric in a steam engine is

(A) equal to the eccentricity
(B) equal to 1/2 the eccentricity
(C) equal to the distance from the center of the main shaft to the center of the eccentric

(D) equal to the diameter of
the eccentric.

43. The angle of advance in a
simple engine is 30°. The
angle between the crank pin
and the eccentric is

(A) 30°   (C) 60°
(B) 120°   (D) 150°.

44. The use of a tail rod for the
same diameter cylinder is more
advisable in a uniflow than a
counterflow engine because in
a uniflow engine

(A) steam admission is in the
center
(B) the m.e.p. is greater
(C) the piston is heavier
and longer
(D) the speed is higher.

45. In boiler feedwater "temporary hardness" is a term
denoting

(A) hardness that quickly
disappears upon exposure
to atmosphere
(B) hardness that can mostly
be precipitated or thrown
down upon application of
heat
(D) hardness that requires
chemicals but no heating
for its removal.

46. The principal means of preventing objectionable odors
from the incinerator is to

(A) keep the incinerator temperature as low as pos-
sible
(B) operate with as much air
as possible
(C) keep the incinerator
temperature as high as
possible
(D) sprinkle lime in the
furnace.

47. A boiler horse power is
equivalent to

(A) 30 lbs. of steam
(B) 2545 Btu per hr.
(C) 34.5 lbs. of steam
evaporated per hr. from
and at 212°F.
(D) 35.5 lbs. of steam
evaporated per hr. from
and at 212°F.

48. In a centrifugal pump

(A) hp. approximately varies
directly with the speed
(B) hp. approximately varies
as the square of the speed
(C) gpm approximately varies
as the square of the speed
(D) head approximately varies
as the square of the speed.

49. In a plumbing system a leader

(A) is the upper part of a
waste stack
(B) connects a soil stack to
the house trap
(C) connects a series of
toilets
(D) conducts the drainage of
the roof to the house
drain.

50. The ordinary 2½-gallon soda
and acid hand fire extinguish-
er in structures used for
amusement or instruction purposes must be refilled

(A) every two years
(B) every year
(C) at the discretion of the
inspector
(D) only after emptying in use.

51. In a single pressure stage,
two-velocity stage, impulse
steam turbine operating on 175
lbs. gauge pressure with 5
lbs. gauge back (exhaust)
pressure, the pressure of the
steam between the first
moving row and the stationary
row of blading is most nearly

(A) 5 lbs. gauge
(B) 175 lbs. gauge

(C) 170 lbs. gauge

(D) 85 lbs. gauge.

52. In a two-pipe heating system the use of radiator inlet orifices usually

(A) necessitates the use of a vacuum of 15" or more

(B) tends to equalize distribution to radiators and give higher radiator temperatures

(C) is economical because they permit the use of higher vacuum

(D) saves steam by permitting lower radiator temperatures with equal distribution.

53. When fired with the same percentage of excess air, for the same boiler horsepower developed, fuel oil compared to coal, has

(A) less weight, more bulk, lower CO_2

(B) less weight, less bulk, lower CO_2

(C) more weight, less bulk, lower CO_2

(D) less weight, more bulk, higher CO_2.

54. Assume that the principal elements of a refrigeration system are (1) cooler (2) liquid receiver (3) compressor (4) expansion valve (5) oil separator (6) condenser. The order in which these elements are usually arranged is

(A) 4-1-3-5-6-2

(B) 5-1-6-3-2-4

(C) 2-1-4-5-3-6

(D) 3-4-5-2-1-6.

55. A pneumatic sewage ejector is

(A) a device for pumping air from the building plumbing system

(B) a pump for forcing out obstructions in plumbing systems

(C) an automatic syphoning trap for plumbing fixtures

(D) a device for forcing sewage by compressed air from a lower level to a level where it can enter the sewer.

56. In a shell-and-coil hot water heater with inlet and outlet shut-off valves on the water side, the one of the following that is most important, if the coil is supplied 10 lb. steam, is a

(A) 2" asbestos covering on the tank

(B) thermostatic control valve

(C) relief valve on the tank

(D) relief valve on the steam supply.

57. The purpose of a barometric damper is to

(A) maintain constant draft loss across the boiler

(B) reduce smoke

(C) increase draft loss across the ash pit

(D) decrease CO_2 at boiler outlet.

58. When two or more high pressure boilers (over 15 lbs. pressure) are connected to a common steam header, the A.S.M.E. Code requires

(A) a valved outlet in the boiler outlet between the automatic stop valve and the header valve

(B) a relief valve in the main steam header of a sufficient size to relieve the product of either boiler

(C) a by-pass around the low pressure reducing valve at least 1/2 the size of the valve

(D) only one safety valve per boiler (up to 125 hp.).

59. In making up a belt-and-spigot cast iron soil or vent joint, the proper procedure is

   (A) pour metal, caulk inside edge, place oakum, tamp, pour metal, caulk
   (B) place oakum, caulk down, pour metal
   (C) place gasket, place oakum, pack down, pour metal
   (D) place oakum, tamp down, pour metal, caulk.

60. In the absorption system of refrigeration steam is used

   (A) to effect the thermal circulation and no pump is required
   (B) to drive off the ammonia gas from the liquor
   (C) to cause the absorption of ammonia in the liquor
   (D) in the rectifier to rectify the ammonia.

61. The terms "Tuyers and retort" as applied to coal burning are, respectively,

   (A) the throat and the rear part of a chain grate stoker furnace
   (B) the space receiving the coal under the fire and the part of the fire where the volatiles are distilled off
   (C) the pusher that spreads the coal and the hopper receiving the coal in the furnace
   (D) any air nozzle and the air chamber before the nozzle.

62. A radiator is not a convector type unless it

   (A) is hung from ceiling as compared to one supported by legs on the floor
   (B) has a fan to force air across extended surface tubes

   (C) transfers heat almost entirely by heating air flowing through it
   (D) transfers heat primarily by radiation to convected air.

63. The steam lap on a slide valve engine is

   (A) width of port minus the exhaust lap
   (B) the distance that the valve when in its mid-position extends beyond the edge of the steam port towards that side from which it takes steam
   (C) the distance that the valve must travel from the end of its stroke before steam is admitted
   (D) the distance that the valve must travel when the piston is at the end of its stroke before steam is admitted.

64. If you are burning 100 tons of coal per year and your boiler efficiency is 50%, how many tons would you burn if you increased the boiler efficiency to 60%?

   (A) 90   (C) 80
   (B) 120   (D) 66-2/3.

65. In installing a water closet

   (A) the threaded joint of the fixture to the bend is packed with oakum thread
   (B) the china bottom is bolted to the flange of the bend with a gasket
   (C) the spigot joint of the china is fixed into the bell of the bend with oakum and poured metal
   (D) the joint between the china bottom and the bend is held watertight by a threaded compression retainer.

66. A pressure reducing valve reduces steam pressure from 150 lbs. gauge, dry and saturated, to 25 lbs. gauge. Neglecting any radiation from the valve body, this steam at 25 lb. pressure

    (A) has less heat than at 150 lbs.
    (B) has more heat than at 150 lbs.
    (C) has same heat as at 150 lbs.
    (D) is dry and saturated.

67. In figuring the theoretical horsepower of a pump the measured height of the discharge is 100 ft. above the center line of the pump, the water is 20 ft. below the center line of the pump, and friction is taken as 10 ft. What is the total head on the pump?

    (A) 100 ft.
    (B) 56.3 lbs. per sq. in.
    (C) 118 ft.
    (D) 110 ft.

68. Neglecting header and drum surface, a boiler having 144 tubes, 4" diameter, 18 ft. long, would be rated in hp. at approximately

    (A) 700   (C) 275
    (B) 174   (D) 375.

69. The total pressure of a gas flowing in a duct at any point is

    (A) the sum of velocity and static pressures
    (B) the static pressure plus the barometric pressure
    (C) the velocity pressure less the static pressure
    (D) the velocity pressure plus the static pressure plus the friction pressure loss.

70. In two-to-one roping of elevator cables

    (A) the surface speed of the drum is half that of the elevator
    (B) the drum surface speed is twice that of the elevator
    (C) the elevator and counterweight ropes are wound on different drums
    (D) there are twice as many elevator cables as counterweight cables.

71. An ammonia refrigeration system is best purged of air

    (A) at the top of condenser
    (B) at the suction of the compressor
    (C) at the oil separator
    (D) at the liquid receiver.

72. Saybolt refers to a measure of

    (A) specific gravity
    (B) viscosity
    (C) fire clay melting point
    (D) diesel fuel knock rating.

73. Degrees A.P.I. refers to

    (A) Btu or heat content
    (B) temperature
    (C) ignitability or flash point
    (D) density.

74. In a "split" system

    (A) the fixtures above sewer level drain by gravity while the lower ones require a pump
    (B) radiators are used for heat transmission losses and fans for air change
    (C) steam risers for radiation are partly up feed and partly down feed
    (D) the burners are so designed as to burn either oil or pulverized coal.

75. Adding 200 F. superheat to steam decreases the steam consumption of a turbine by 18%. It requires 10% more fuel per pound of steam to add the

superheat. The net result,
in terms of the amount of
fuel used for saturated
steam, is

(A) 8% saving in fuel
(B) 90.2% as much fuel
(C) 82% as much fuel
(D) 17.6% saving in fuel.

Answer Key

1.	D	16.	D	31.	C	46.	C	61.	E
2.	B	17.	C	32.	C	47.	C	62.	C
3.	B	18.	D	33.	B	48.	D	63.	B
4.	A	19.	C	34.	D	49.	D	64.	C
5.	D	20.	C	35.	B	50.	B	65.	B
6.	B	21.	B	36.	A	51.	A	66.	C
7.	C	22.	C	37.	B	52.	D	67.	B
8.	A	23.	B	38.	A	53.	B	68.	C
9.	B	24.	C	39.	D	54.	A	69.	A
10.	D	25.	D	40.	D	55.	D	70.	B
11.	D	26.	D	41.	A	56.	C	71.	A
12.	A	27.	D	42.	A, C, D	57.	D	72.	B
13.	A	28.	C	43.	B	58.	A	73.	D
14.	B	29.	D	44.	C	59.	D	74.	B
15.	A	30.	A	45.	B	60.	B	75.	B

SAMPLE PRACTICE EXAMINATION 9

HIGH PRESSURE PLANT TENDER

DIRECTIONS: Each question has four suggested answers, lettered A, B, C, and D. Decide which one is the best answer, locate the question number on the sample answer sheet, and with a soft pencil darken the area that corresponds to the answer you have selected. All sample answer sheets follow page 22.

The time allowed for the entire examination is 3 hours.

1. Clinkering of coal in a furnace is due mainly to

   (A) too high a percentage of ash in the coal
   (B) too low a fusion temperature of ash
   (C) too high a fusion temperature of ash
   (D) too low a percentage of ash in the coal.

2. The boiler drum of a high pressure boiler is usually constructed with a

   (A) lap joint
   (B) butt and strap joint
   (C) thermit welded joint
   (D) spot welded joint.

3. In a properly designed turbine admitting steam at 100 psi the one of the following sets of steam conditions which would result in the lowest water rate is

   (A) saturated steam with 5 lb. back pressure

   (B) saturated steam with 29 in. vacuum
   (C) 200° superheat with 5 lb. back pressure
   (D) 200° superheat with 29 in. vacuum

4. The heat exchanger between the low pressure and high pressure cylinders of an air compressor is used to

   (A) increase the air pressure
   (B) lower the air temperature
   (C) lower the relative humidity
   (D) raise the temperature of the air.

5. The automatic purger is used on an ammonia refrigerating system to

   (A) remove non-condensible gases from the system
   (B) release excess ammonia pressure
   (C) control suction pressure
   (D) separate oil from compressed ammonia gas.

6. In a reciprocating steam pump

with 200 lb. steam pressure, 6" diameter steam piston, and 5" diameter water piston, the resulting water pressure should be most nearly, in lbs. per sq. in.,

(A) 390   (C) 190
(B) 290   (D) 90.

7. The usual activating element in a thermostatic trap is

(A) a bi-metallic contactor
(B) a float
(C) a flexible metallic bellows
(D) a column of mercury.

8. The one of the following that is needed to determine the volume of air flowing through a duct is

(A) velocity pressure
(B) static pressure
(C) total pressure
(D) negative pressure.

9. The variable cutoff type of governor on a steam engine controls the

(A) size of the exhaust port opening
(B) timing of the steam port opening
(C) steam pressure
(D) size of the steam port opening.

10. In a jet condenser

(A) circulating water flows through the tubes, and steam is on the outside of the tubes
(B) steam is in the tubes, and circulating water is on the outside of the tubes
(C) steam mixes directly with circulating water
(D) ammonia mixes directly with circulating water.

11. The percent of volatile com-

bustible matter in a soft coal from the west Pennsylvania high volatile field would be most nearly

(A) 50%   (C) 70%
(B) 90%   (D) 30%.

12. Oxygen in boiler feedwater causes

(A) priming   (C) scaling
(B) foaming   (D) pitting.

13. If a steam gauge registers 150 lbs. per sq. in. and a thermometer in the steam reads 500°F., the steam is most likely

(A) dry saturated
(B) superheated
(C) wet
(D) none of the above.

14. The CO_2 in the flue gas is 14% over the fire, 11% in the second pass, and 9% at the damper. This most likely indicates

(A) leaks through side walls
(B) too much air under the fire
(C) incomplete combustion
(D) not enough air under the fire.

15. In a three-wire, 240 volt D.C. lighting system using 120 volt lamps, one line carries 250 amperes when another line carries 150 amperes. This indicates

(A) an open circuit
(B) a short circuit
(C) a fault to ground
(D) none of the above.

16. The one of the following substances not commonly used as a refrigerant is

(A) methyl chloride
(B) sodium fluoride
(C) sulphur dioxide
(D) carbon dioxide.

17. Shrouds are most likely to be found in

    (A) steam boilers
    (B) condensers
    (C) oil coolers
    (D) steam turbines.

18. In relation to the setting of the boiler safety valves the safety valves on the super-heater are usually set

    (A) lower
    (B) higher
    (C) the same
    (D) none of the above because there are no safety valves on the superheater.

19. The best practice in heating heavy fuel oil is to use as the heating medium

    (A) high pressure steam
    (B) an oil burner
    (C) a gas burner
    (D) hot water.

20. A three-wheel pipe cutter is preferable to a one-wheel pipe cutter when the pipe to be cut is

    (A) less than 1" in size
    (B) longer than 6' in length
    (C) a 2" riser in the corner of a room
    (D) held in position by a pipe vise.

21. To measure relative humidity, use is made of

    (A) one dry bulb thermometer
    (B) one wet bulb thermometer
    (C) one wet bulb thermometer and one dry bulb thermometer
    (D) two wet bulb thermometers.

22. In starting a centrifugal boiler feed pump with 300 lb. water pressure on the line, the valves should be set with

    (A) both suction and discharge open
    (B) suction open, discharge open
    (C) both suction and discharge closed
    (D) suction closed, discharge open.

23. The temperature of flue gas leaving a straight water tube boiler increases suddenly and remains high although there was no increase in rating. This most likely indicates failure of the

    (A) stoker   (C) baffle
    (B) damper   (D) fan.

24. The one of the following names used to designate a commer-cially available size of soft coal is

    (A) #3 Buckwheat
    (B) Breeze
    (C) Nut-slack
    (D) Nonpareil.

25. When dry saturated steam at 150 lbs. per sq. in. pressure is reduced to 10 lbs. per sq. in. through a reducing valve, the low pressure steam would

    (A) be superheated
    (B) be dry saturated
    (C) have 5% moisture
    (D) have 10% moisture.

26. Early cutoff in a steam engine is most likely to cause

    (A) detonation
    (B) condensation
    (C) reversal of engine direc-tion
    (D) superheating.

27. A shear pin would be most likely found in

    (A) a stoker
    (B) a centrifugal pump
    (C) a duplex steam pump
    (D) a motor-generator set.

28. In a gearless traction electric elevator, the brake is

(A) mounted on the governor
(B) set manually from the car
(C) mounted under the car
(D) mounted on the same shaft as the sheave.

29. A reversing diesel engine most likely will have

(A) a Walschaert gear
(B) four sets of poppet valves
(C) two camshafts
(D) a large flywheel.

30. The best time to blow down a boiler is

(A) after soot blowing
(B) before soot blowing
(C) before banking
(D) after banking.

31. A weir is used to measure the flow of

(A) steam   (C) water
(B) air   (D) gas.

32. A freon (F-12) leak is best located by means of

(A) sulphur candle
(B) tincture of green soap
(C) a halide torch
(D) litmus paper.

33. In a mechanical type of oil burner the atomization of the oil is produced by the

(A) steam pressure
(B) burner tip
(C) strainer
(D) oil heater.

34. The most efficient method of controlling the volume of air delivered by a fan is

(A) control of outlet damper
(B) control of intake damper
(C) varying the fan speed
(D) by-passing the fan.

35. In a Sterling or bent tube type of boiler the superheater is usually located in the

(A) first pass
(B) second pass
(C) third pass
(D) uptake.

36. In an electric elevator the car gate interlock switch is wired in series with the

(A) motor armature
(B) corridor door interlock switches
(C) motor field
(D) accelerating resistor.

37. A deaerating heater is used on

(A) feedwater
(B) fuel oil
(C) lubricating oil
(D) absorption refrigerating systems.

38. Excessive air in a shell and tube steam condenser causes

(A) decreased vacuum
(B) scaling
(C) increased vacuum
(D) excess cooling water.

39. The term SAE used to describe lubricating oil numbers refers to

(A) Standard Auto Engine
(B) Safe and Approved for Engines
(C) Bureau of Standards and Engineering
(D) Society of Automotive Engineers.

40. In a steam plant the flue gases usually pass through the (1) boiler, (2) induced fan, (3) air heater, (4) forced fan, and (5) economizer, in the following order:

(A) 2-1-4-3-5   (C) 2-1-5-3-4
(B) 4-1-5-3-2   (D) 4-3-5-1-2.

41. The number of sets of valves in the usual type of reciprocating wet vacuum pump is

(A) 2   (C) 4
(B) 1   (D) 3.

42. For winter operation an evaporative condenser is frequently operated as

(A) a vertical shell and tube condenser
(B) a double pipe condenser
(C) an atmospheric condenser
(D) a horizontal shell and tube condenser.

43. In a fan

(A) hp. varies directly with the speed
(B) volume varies as the cube of the speed
(C) hp. varies as the cube of the speed
(D) draft varies directly with the speed.

44. Plumbers "soil" is used in making up joints in

(A) lead pipe
(B) cast iron pipe
(C) wrought iron pipe
(D) copper tubing.

45. The holes in the firedoor of a hand-fired furnace

(A) should be kept closed
(B) are provided to prevent the door from warping
(C) allow secondary air to enter the furnace
(D) should be opened wide if the boiler setting is very leaky.

46. The percent of total steam generated by the boiler that is required to operate a steam atomizing oil burner is most nearly

(A) 5%   (C) 10%
(B) 1%   (D) 25%.

47. An automatic non-return valve is used on a

(A) turbine
(B) condenser
(C) pump
(D) boiler.

48. A spanner wrench is used on an ammonia compressor to

(A) regulate the pump-out valve
(B) make up on the stuffing box gland
(C) back-seat suction and discharge valves
(D) rotate the eccentric.

49. The published boiler code under the provisions of which boilers in New York City are inspected is that of the

(A) U.S. Bureau of Standards
(B) Department of Buildings
(C) Middle Atlantic Boiler Inspectors Conference
(D) New York State Department of Labor.

50. Suppose that a boiler has an efficiency of 80%, and is fired with a fuel whose heating value is 1400 Btu per lb. Each 10 ft. length of uninsulated steam pipe in this plant loses 2,000 Btu per hr. Insulation can reduce this loss to 200 Btu per hr. Approximately how many lbs. of fuel per hour can be saved by insulating all 100 ft. of steam piping in this plant?

(A) 10 lbs/hr.   (C) 16 lbs/hr.
(B) 13 lbs/hr.   (D) 20 lbs/hr.

Answer Key

1.	B	11.	D	21.	C	31.	C	41.	D
2.	B	12.	D	22.	B	32.	C	42.	C
3.	D	13.	B	23.	C	33.	B	43.	C
4.	B	14.	A	24.	C	34.	C	44.	A
5.	A	15.	D	25.	A	35.	A	45.	C
6.	B	16.	B	26.	B	36.	B	46.	A
7.	C	17.	D	27.	A	37.	A	47.	D
8.	A	18.	A	28.	D	38.	A	48.	B
9.	B	19.	D	29.	C	39.	D	49.	B
10.	C	20.	C	30.	D	40.	B	50.	C

SAMPLE PRACTICE EXAMINATION 10

SHORT ESSAY TYPE EXAMINATION

High Pressure Plant Tender

DIRECTIONS: Answer these questions on a separate piece of blank paper. Answer them completely but use a minimum amount of words to do so. Then compare your answers with those given at the end of the examination.

1. List the sizes of anthracite coal which are commonly used in power plant work.

2. List the sizes of bituminous coal which are commonly used in power plant work.

3. Give some of the good features of the horizontal return tubular boiler.

4. Give some of the good features of water tube boilers.

5. A. What is meant by priming?

   B. What is the cause of priming?

6. A. How does scale in a boiler affect it when not enough is present to be dangerous?

   B. How can the accumulation of scale in a boiler be reduced and kept at a minimum?

7. A. What should be done to the safety valves on a high pressure steam boiler when preparing for a hydrostatic test?

   B. How would you prepare a boiler for inspection?

8. What is the first duty of a stationary fireman upon taking over a watch in a high pressure steam plant?

9. A. In what units are draft pressures generally indicated?

   B. In what units are steam pressures generally indicated?

10. A. In a high pressure steam plant what devices and means are used to remove soot from the tubes of a water tube boiler?

    B. What is meant by turbining tubes?

11. A. What is a "dead plate" and with what specific kind of fuel is it used?

    B. How does soft coal differ

from hard coal, aside from differences in size?

12. A. How many and what type of valves are generally found in a blowdown line on a high pressure boiler?

B. What is the reason for blowing down a boiler?

C. When should a boiler be blown down?

13. A. What is a fair CO_2% value for a steam boiler using hard coal as a fuel?

B. What is a fair CO_2% value for a steam boiler using No. 6 fuel oil as a fuel?

C. A flue gas analysis showed presence of CO. What does this mean to you in regards to combustion efficiency?

14. A. If you came on watch and noticed the water level is not visible in the gauge glass, what would you do?

B. On a high pressure steam boiler where should the try-cocks be located, and for what are they generally used?

C. On high pressure boilers what is the purpose of the water column?

15. A. How would you regulate the speed of a duplex steam driven boiler feed pump?

B. What lubrication is required on the water end of an outside packed, plunger type, steam driven, duplex feedwater pump?

16. A. In a high pressure steam boiler name the order in which draft will vary from the highest positive pressure to the lowest negative

pressure.

B. What is the purpose of a balanced draft device on a high pressure steam boiler?

C. In an oil-fired steam boiler what is the primary purpose of a barometric damper?

17. A. List the grades of fuel oil and classify them according to use.

B. At what pressure and temperature should a No. 6 fuel oil be delivered to the cup of a horizontal rotary cup type oil burner?

C. At what pressure and temperature should a No. 6 fuel oil be delivered to the gun of a mechanical pressure type oil burner?

18. A. Give the definition of a boiler horsepower.

B. How many pounds of steam per hour, from and at 212°F., is generated by a 100 horsepower boiler operating at 100% rating?

19. A. What is the purpose of a checkerboard floor construction in the furnace of a No. 6 oil rotary cup type burner?

B. The flame in the combustion chamber of a boiler using a mechanical pressure type burner smokes badly. What is most likely the trouble?

20. A. In an oil-fired installation what is the purpose of the Protectostat?

B. In a horizontal rotary cup type burner installation, with what other control should the Vaporstat be electrically interlocked?

21. What equipment should you constantly use to reduce the possibility of injury to your person when at work in an incinerator plant?

22. What is the basic difference between an incinerator furnace and the furnace of a hand-fired coal-burning water tube boiler?

23. In an incinerator furnace where does slagging take place?

24. In a coal-fired burner where does slagging take place?

25. What is fly-ash?

26. How is slag removed from an incinerator furnace?

27. What type of water tube boiler is equipped with stay bolts? Where are these stay bolts located?

28. List the different kinds of bars which are commonly used in stoking incinerator fires.

29. Why is it essential to stoke incinerator fires?

30. What is a "lazy" bar and when should it be used?

31. For an incinerator furnace equipped with forced draft, what should the draft pressure be in the ash pit for normal operating conditions?

32. For an incinerator furnace equipped with forced draft, if the ash pit draft pressure is too low, what would you do?

33. What type of instrument is generally used to indicate furnace temperature?

34. With respect to a fire tube boiler what is meant by the phrase "punching the tubes?"

35. What means or "tricks of the trade" can you use to check the combustion efficiency of a coal-fired boiler?

36. In an arrangement of incinerator furnaces and waste heat boilers, for what is a by-pass gate used?

37. In an incinerator plant what care must be practiced when using an air preheater along with a forced draft system?

38. Briefly describe how you would bank a coal fire in a boiler operating at 50 lbs. per sq. in. steam pressure.

39. Of what value is a CO_2 analysis taken on a steam generated boiler using No. 6 fuel oil?

40. Briefly, what is meant by the "coking" method of hand firing in a furnace?

41. What is the basic difference between a shaking grate and a dumping grate?

42. A steam generating boiler is equipped with two blowdown lines. State briefly just how you would go about blowing down this unit.

43. A boiler is one of a bank of units. Each boiler is equipped with an inside and outside stop valve on the boiler steam branch line. A blowdown is provided between these two stop valves. State when you would use this blowdown line.

44. What devices or means are generally provided to keep firing doors from burning out?

45. Briefly state why it is necessary to chemically treat feed water in a steam generating plant.

46. When packing the steam end of a duplex feedwater pump, what care must be taken in placing the packing rings in position into the packing gland?

47. What care must be taken when filling a boiler before firing it in order to place it on the line?

48. With regard to a coal hand-fired boiler what is meant by the term "char"?

49. What would be the efficiency of a boiler using anthracite coal having a calorific value of 13,200 Btu per lb.? The heat content of the feedwater increases 1,050 Btu in changing to steam. The boiler produces 10 lbs. of steam per lb. of coal.

50. Indicator cards taken from an 11" x 13" double-acting steam engine running at 200 rpm each have an area of 1.80 square in. and a card length of 3". An 80 spring was used in the indicator. Neglecting the area of the piston rod, calculate the indicated horsepower.

51. What hp. is required to operate a boiler feed pump taking water from a heater at 205 lbs. per sq. in. and discharging to a boiler at 625 lbs. per sq. in.? The friction drop in the discharge line is 50 lbs. per sq. in. The pump efficiency is 87% and the water flow is 254,000 lbs. per hr. One lb. per sq. in. equals 2.34 ft. of water.

52. A turbine exhausts 148,000 lbs. of steam per hr. with a heat content of 1080 Btu per lb. to a condenser at 1.5 in. of mercury pressure. How many gal. per min. of condensing water are required if the water enters at 60°F. and leaves at 73°F.? The heat content of the condensate leaving the condenser is 59 Btu per lb. (One gal. of water weighs 8 1/3 lbs.)

END OF WRITTEN TEST

Explanatory Answers

1. Steam power plants usually burn the smaller, less expensive sizes of anthracite. The trade names for the larger to the smaller sizes are: Pea, Buckwheat No. 1, Buckwheat No. 2, Buckwheat No. 3, Rice, Barley, and Culm.

2. The sizes of bituminous coal usually used in steam power plants are: Run-of-the-mine, Slack, and Culm.

3. They are compact, have water storage capability, and a lower initial cost.

4. Water tube boilers respond more readily to heavy and sudden demands for high quality steam, particularly at high pressure. They are easier to clean and maintain.

5. A. Priming is the discharge of slugs of water along with the steam. It causes dirt to be carried by the steam to foul traps, steam lines, superheaters, and turbines.
   B. Priming is caused by boiling water being thrown up into the steam space of boilers having inadequate steam and water disengaging surfaces, or boilers that are carrying too high a water level. Oil on boiler surfaces may also contribute to priming.

6. A. Scale insulates the metal surfaces, and therefore reduces boiler capacity. It may also cause overheating and weakening of the boiler tubes and shell plates.
   B. Frequent blowdowns and proper chemical water treatment will reduce scale accumulation and formation.

7. A. Install gags on safety valves after hydrostatic pressure has caused the safety valves to lift.
   B. Take boiler off the line, dump water, clean water side, open manholes and handholes, and clean flue side.

8. Blow down the water columns. Check boiler water levels. Ask if there are any unusual problems.

9. A. Drafts are indicated in positive or negative equivalent inches of water.
   B. Steam pressures are indicated in lbs. per sq. in. above atmospheric pressure (psi g).

10. A. Mechanical soot blowers using jets of air or a mixture of steam and air.
    B. Removing scale from inside the water tubes by means of mechanical tube cleaners using revolving cutters driven by compressed air, steam, or water under pressure.

11. A. The "dead plate" is the front, fixed portion of the fire box on a coal burning boiler.
    B. Soft coal burns with a long yellow smokey flame. It has a higher percentage of volatile matter and sulphur.

12. A. Two. The valve nearest the boiler is usually a seatless plunger type which the operator opens last and closes first when blowing down. The other valve is usually either a quarter turn cock and shutter valve, or a seat and disk valve.
    B. To rid mud drums of their accumulations, to drain the boiler, to minimize scale formation by deconcentration, and to lower boiler water levels.
    C. Whenever the boiler has to be drained, or is over-filled. Some boilers have a continuous blowdown provision for deconcentration of the boiler water. Priming and foaming are indications that blowdowns are required. Blowdowns should be coordinated with anticipated load changes and water treatment.

13. A. 16%
    B. 13%
    C. Reduced combustion efficiency due to incomplete combustion.

14. A. Blow down the water column. Use the try-cocks to determine whether the boiler is over or under-filled. Take the appropriate action to either blow down or shut down.
    B. Try-cocks are located near the top, middle and bottom of the water column gauge glass. Under normal operation the top try-cock should produce steam, the middle one steam and water, the bottom one water only.
    C. The water column is connected to the front of the boiler drum with its top connected to the steam space and its bottom to the water space. It is designed to normally indicate at its midpoint. It provides visual indication that the boiler water level is correct for firing.

15. A. Throttling the steam inlet by adjusting the throttling governor.
    B. Graphite and oil.

16. A. Forced draft fans blow air into closed boiler rooms so that air under the highest pressure is forced into the furnace under and through the fuel bed. The products of combustion are removed through the stack by natural, or induced drafts, with the lowest pressure.
    B. Balanced draft devices control forced and induced drafts to produce practically atmospheric pressure in the furnace over the fire.
    C. The barometric damper is a draft regulating device located in the boiler's smokepipe to control the over-the-fire draft. It can also close the smoke outlet when the burner is idling in order to conserve heat.

17. A. Fuel oil grades run from the lightest - No. 1 through Nos. 2, 3, 4, and 5, to the heaviest - No. 6. Diesel oils are designated as Nos. 1-D, 2-D, etc. No. 2 fuel oil is used without preheat in domestic type oil burners. No. 6 fuel oil requires preheating, and is usually used in high pressure plants.
    B. No. 6 fuel oil should be supplied to horizontal rotary cup type oil burners at temperatures and pressures of approximately 100°F. and 5 to 40 psi.
    C. No. 6 fuel oil should be supplied to mechanical pressure type oil burners at temperatures and pressures of approximately 100°F. and 200 psi.

18. A. One boiler horsepower is equivalent to the evaporation of 34.5 lbs. of water per hr. at atmospheric

pressure and a temperature of 212°F.

B. 3,450 lbs. per hr.

19. A. To protect furnace water walls or refractory lining.
    B. Insufficient atomization due to low oil pressure or temperature.

20. A. Shut down, or prevent firing when pressures, water levels or other conditions are unsafe.
    B. Ignition, magnetic fuel valves, burner motor, puge cycle controller.

21. Hard-hat, safety shoes, goggles with smoked glass lenses.

22. Some incinerators are not designed to generate process steam, and therefore lack heat transfer or recovery arrangements. Incinerators usually have moving grates.

23. Against the walls.

24. On the grates.

25. Fine air-borne ash carried along by the flue gasses.

26. Replace the affected refractory brick lining.

27. Pulverized coal-burning water tube boilers frequently have stud-tube water walls covered with cast iron blocks held to the tubes by stay bolts.

28. Long iron rakes, hoes and push bars.

29. Improve the combustion process and reduce the amount of unburned refuse.

30. To cut hot slag.

31. Positive pressure of about 0.10" of water.

32. Check forced draft fans.

33. Pyrometer.

34. Cleaning soot from the tubes.

35. Observe the flames. They should be a barely visible blue if anthracite is being burned. Check for dense black smoke.

36. When process steam from waste heat recovery is not required.

37. Preheating the air may cause furnace temperatures to become excessive.

38. Slant the newly added coal up towards the rear and sides of the firebox. Rake the burning coals towards the feed door. Create a "hot spot" of exposed, red glowing coals. Adjust primary and secondary air for proper combustion.

39. Indicates combustion efficiency.

40. Coking grade bituminous coal fired in a deeper than usual bed to drive off volatile components.

41. Shaker and dumping grates are both built of movable sections operated from the front of the furnace by hand levers. Shaker grates can be rocked back and forth by the lever action.

42. One line provides mud drum continuous blowdown for deconcentration of boiler water, and is automatically controlled by a pressure regulator. The other line is a bottom blowdown which is manually operated by means of a seatless and seated valve operating in proper sequence, as required to minimize fouling and scale.

43. Drain condensate from the steam lines after the boiler has been removed from the line and before placing it on the line.

44. Water cooling the combustion chamber.

45. Prevents corrosion, fouling of the heat absorbing surfaces and contamination of the steam.

46. Packing rings should be of correct size to prevent distortion or binding and to make a steam tight installation.

47. Boiler should be cold and water properly conditioned.

48. Partially burned coal.

49. Efficiency equals

$$\frac{10}{1} \times \frac{1050}{13,200} \times 100 = 79.6\%$$

50. m.e.p. equals $\dfrac{80 \times 1.80}{3}$

$$= 48 \text{ psi.}$$

Ihp. equals $\dfrac{2(PALN)}{33000}$ =

$$\frac{2(48 \times \pi \times 11 \times 11 \times 13 \times 200)}{(33000 \times 4 \times 12)} =$$

59.8

51. Pump hp. =
$$\frac{254,000 \ (625.205 + 50) \times 2.34}{0.87 \times 33,000 \times 60}$$
$$= 162 \text{ hp.}$$

52. gpm = $\dfrac{148,000 \times (1,080 - 59)}{60 \ (73-60) \times 8}$

$$= 23,246$$

SAMPLE PRACTICE
EXAMINATION 11

STATIONARY ENGINEER (ELECTRIC)

DIRECTIONS: Each question has four suggested answers, lettered A, B, C, and D. Decide which one is the best answer, locate the question number on the sample answer sheet, and with a soft pencil darken the area that corresponds to the answer you have selected. All sample answer sheets follow page 22.

The time allowed for the entire examination is 3½ hours.

1. The device which is most commonly used to measure the insulation resistance of an electrical circuit is a

   (A) growler   (C) megger
   (B) bolometer   (D) wattmeter.

2. A wire 60 feet long is divided into two parts with the lengths of the respective parts in the ratio of 1 to 3. Under these conditions the length of the shorter piece is

   (A) 5'   (C) 15'
   (B) 10'   (D) 20'.

3. Silver electrical contactors are tarnished most readily by

   (A) oxygen   (C) nitrogen
   (B) hydrogen (D) sulphur.

4. The one of the following devices which is commonly used to prevent damage in cases of reversal of leads in reconnecting the wiring of three-phase motors is a

   (A) reverse current relay
   (B) reverse power relay
   (C) reverse phase relay
   (D) reverse power factor relay.

5. The two wattmeter method is used to measure the power delivered to a three-phase balanced delta connected load. If one of the two wattmeters used reads zero, the power factor of the load is most nearly

   (A) 1   (C) .5
   (B) .8   (D) 0.

6. The devices most commonly used to limit the current on short circuit in high tension feeder circuits are

   (A) resistors (C) capacitors
   (B) reactors   (D) resonators.

7. The purpose of laminating the core of a power transformer is to keep the

   (A) hysteresis loss at a minimum

(B) eddy current loss at a minimum
(C) copper losses at a minimum
(D) friction losses at a minimum.

8. In a given circuit when the power factor is unity, the reactive power is

(A) a maximum
(B) zero
(C) equal to I^2R
(D) a negative quantity.

9. The winding pitch for a 2-pole lap wound D.C. armature having 22 shots is approximately

(A) 44   (C) 22
(B) 33   (D) 11.

10. The no-load speed of a 6-pole 3-phase 60-cycle squirrel cage induction motor is most nearly

(A) 600   (C) 2395
(B) 1198   (D) 3600.

11. The input to a motor is 16,000 watts and the motor losses total 3,000 watts. The efficiency of the motor is most nearly

(A) 68.4%   (C) 84.21%
(B) 81.25%   (D) 87.5%.

12. The weight of a 10 ft. section of a 2" x 1/8" copper bus bar that is made of copper having a density of 0.32 lbs. per cu. in. is most nearly

(A) 0.82 lbs.   (C) 9.6 lbs.
(B) 7.7 lbs.   (D) 30.0 lbs.

13. The armature of a D.C. motor runs hot in spots and cool in others. This trouble is most probably caused by

(A) a short circuited coil or coils
(B) an overload
(C) moisture in coils
(D) the armature being off center between poles.

14. The armature of a D.C. motor runs hot all over. This trouble is most probably caused by

(A) a short circuited coil or coils
(B) open circuited coils
(C) reverse polarity of an armature coil
(D) overload or brown out.

15. The characteristic of a lubricating oil which is indicated by its S.A.E. number is its

(A) flash point   (C) density
(B) viscosity   (D) osmosis.

16. Accurate resistances the values of which are not materially affected by changes in room temperature are usually made of an alloy commonly called

(A) manganin   (C) Excellin
(B) paganin   (D) Siemens Martin.

17. The gears most commonly used to connect two shafts which intersect are usually a form of

(A) spur gears
(B) bevel gears
(C) spiral gears
(D) herringbone gears.

18. A standard No. 8 machine screw thread has

(A) 24 threads per inch
(B) 28 threads per inch
(C) 30 threads per inch
(D) 32 threads per inch.

19. To safely secure a ½" wire rope around a thimble, the minimum number of Crosby or U-type clips which should be used is

(A) 2   (C) 4
(B) 3   (D) 5.

20. The speed of a centrifugal pump driven by a certain induction motor is increased 10%. The quantity of water pumped will then be increased by approximately

(A) 5%   (C) 15%
(B) 10%   (D) 20%.

21. A centrifugal pump is to be
started, operated, and then
stopped. In order to do this,
with respect to pump operation
and valve position, the correct procedure is to

(A) close the discharge valve;
bring up to speed, then
open the discharge valve;
turn off the prime mover
and then shut the dis-
charge valve.
(B) close the discharge valve;
bring pump to speed, then
gradually open the dis-
charge valve; shut the discharge valve and then turn
off the prime mover
(C) open the discharge valve,
then bring up to speed;
gradually close the dis-
charge valve and turn off
the prime mover
(D) open the discharge valve,
then bring up to speed;
close the discharge valve
and turn off the prime
mover.

22. Knife edges which are parallel
and level are often used to
check the impeller of small
pumps for

(A) static balance
(B) dynamic balance
(C) centrifugal balance
(D) reverberation balance.

23. An empty bottle weighs 0.5 lbs.
When full of water it weighs
2.5 lbs. When full of mercury
it weighs 27.61 lbs. Therefore the specific gravity of
mercury is approximately

(A) 10.65   (C) 21.30
(B) 13.55   (D) 27.10.

24. Mercury is commonly cleaned by
forcing it through

(A) a fine cloth

(B) a fine copper mesh
(C) a shammy
(D) filter paper.

25. A 110-volt, 5-ampere wattmeter
is connected to a single-phase
circuit by means of a 5:1 current transformer and a 10:1
potential transformer. If the
wattmeter reads 360 watts the
actual power in the main circuit is most nearly

(A) 3,600 watts
(B) 20,800 watts
(C) 18,000 watts
(D) 31,140 watts.

26. Two 100 Kva transformers are
connected in open delta to a
3-phase 4,400 volt bus. The
capacity of this transformer
bank is most nearly

(A) 300 Kva   (C) 170 Kva
(B) 200 Kva   (D) 86 Kva.

27. If a given compressor requires
a full-load torque of 50 lb ft.
and runs at 1800 rpm, the size
(hp.) of the direct-coupled
motor required to drive this
compressor is approximately

(A) 8.5   (C) 21.3
(B) 17.1   (D) 34.2.

28. A bare stranded cable is made
up of 7 strands, each 110 mils
in diameter. The maximum diameter of this cable is most
nearly

(A) 285 mils   (C) 550 mils
(B) 330 mils   (D) 980 mils.

29. A copper bus bar an inch in
diameter has a cross sectional
area of

(A) 3,141,600 C.M.
(B) 1,000,000 C.M.
(C) 785,400 C.M.
(D) 100,000 C.M.

30. If a 3-phase motor connected
to a 208 volt source takes

100 amperes at 90% power factor, then the input power to this motor is most nearly

(A) 18,720 watts
(B) 20,800 watts
(C) 32,400 watts
(D) 56,160 watts.

31. A stranded conductor has 37 strands each 90 mils in diameter. The area in circular mils of this conductor is most nearly

(A) 3,300   (C) 236,000
(B) 123,200   (D) 300,000.

32. A length of wire 1800' long is made up in a coil. If this coil has an average diameter of 6" then the number of turns in the coil is most nearly

(A) 1,000   (C) 1,450
(B) 1,150   (D) 7,200.

33. In a 3-phase 4-wire volt system in which the voltage between lines is 5,500 volts, the voltage to neutral is most nearly

(A) 9,500 volts
(B) 5,500 volts
(C) 3,175 volts
(D) 1,830 volts.

34. When measuring the speed of a D.C. motor by means of a stopwatch and a revolution counter, the stopwatch was started when the revolution counter read 50. At the end of 80 seconds the counter read 50. At the end of 80 seconds the counter read 1,650. The average rpm of the motor during this period was

(A) 200   (C) 1,600
(B) 3,300   (D) 1,700.

Questions 35 to 39 relate to the diagram below:

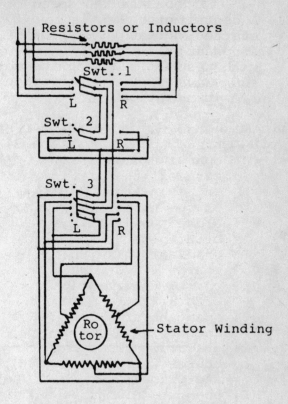

Diagram of a manually operated squirrel cage induction motor.

35. The resistors or inductors are used for

(A) starting the motor
(B) lowering the speed of the motor
(C) reversing the direction of rotation
(D) increasing the speed of the motor.

36. The only purpose for swt. I is to

(A) decrease the speed of the motor
(B) start and stop the motor
(C) reverse the direction of rotation
(D) increase the speed of the motor.

37. The only purpose for swt. II is

    (A) to decrease the speed of
        the motor
    (B) to increase the speed of
        the motor
    (C) to start and stop the motor
    (D) to reverse the direction
        of rotation.

38. If swt. I is in position L and
    switch III is in position R,
    then the motor will run

    (A) at its higher speed
    (B) at its lower speed
    (C) clockwise
    (D) counterclockwise.

39. Swt. III connects the motor
    stator windings in

    (A) delta-delta
    (B) star-star
    (C) such a way that the number
        of poles may be changed
    (D) such a way that the number
        of phases may be changed.

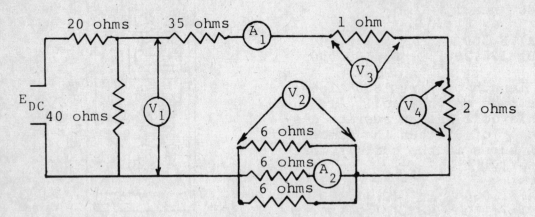

Questions 40 to 47 are based on the diagram above.

40. When the D.C. voltage E_{DC} is
    200 the voltmeter V_1 will read

    (A) 20 volts   (C) 100 volts
    (B) 40 volts   (D) 200 volts.

41. If V_1 reads 120 volts the cur-
    rent reading of ammeter A_1 is
    most nearly

    (A) 2 amperes   (C) 6 amperes
    (B) 3 amperes   (D) 8 amperes.

42. If the reading of ammeter A_1 is
    12 amperes, the reading of ammet-
    er A_2 is most nearly

    (A) 4 amperes   (C) 6 amperes
    (B) 5 amperes   (D) 8 amperes.

43. If the reading of ammeter A_1 is
    2 amperes, the voltmeter V_1 reads
    most nearly

    (A) 36 volts   (C) 12 volts
    (B) 30 volts   (D) 4 volts.

44. If voltmeter V_2 reads 24 volts

the reading of voltmeter V_3
will be most nearly

    (A) 3 volts   (C) 12 volts
    (B) 6 volts   (D) 36 volts.

45. If V_1 reads 120 volts and two
    of the 6-ohm resistors are
    shorted, the ratio of the
    readings of V_3 and V_4 will

    (A) vary inversely as the
        line current
    (B) remain the same
    (C) increase
    (D) decrease.

46. If V_1 reads 120 volts and
    two of the 6-ohm resistors
    are opened, the ratio
    V_2/V_3 will

    (A) vary inversely as the
        line current
    (B) remain the same
    (C) increase
    (D) decrease.

47. If the current through the 1-ohm resistor is 2 amperes then the combined power consumed by the 1-ohm resistor, the 2-ohm resistor and all three 6-ohm resistors is

(A) 10 watts   (C) 42 watts
(B) 20 watts   (D) 84 watts.

48. A pump is required to pump 60 mgd against a 65 ft. head. If the pump efficiency is 65%, the hp. required is most nearly

(A) 680   (C) 1,050
(B) 850   (D) 2,520.

49. When two identical centrifugal pumps are operated in parallel

(A) the head at which they operate together at a given capacity output is double that for a single pump
(B) the capacity at which they operate together at a given head remains the same as that for a single pump
(C) the capacity at which they operate together at a given head is double that for a single pump
(D) the efficiency at which they operate together at a given capacity is the same as for a single pump at that capacity.

50. An ammeter is connected to the secondary of a current transformer with a ratio of 100 to 5. If the ammeter indicates 2.0 amperes the current in the primary circuit will be most nearly

(A) 2.0 amperes
(B) 10.0 amperes
(C) 40.0 amperes
(D) 200 amperes.

51. A 10 to 1 step-down transformer has 44,000 volts on the primary and 4,400 on the secondary. If the taps are changed to reduce the number of turns

in the primary by 2.5%, then the secondary voltage will be most nearly

(A) 4,290 volts
(B) 4,390 volts
(C) 4,410 volts
(D) 4,510 volts.

52. An integrating watthour meter has 4 dials. If from left to right the respective pointers 6 and 7, between 0 and 1, and between 4 and 5, then the reading is

(A) 9715   (C) 4179
(B) 8604   (D) 4068.

53. In an A.C. circuit a low value of reactive volt amperes compared with the watts indicates

(A) maximum current for the load
(B) low efficiency
(C) high power factor
(D) unity power factor.

54. A 50 millivolt meter shunt having a resistance of 0.005 ohms is to be used with a 5 milliampere meter whose resistance is 10 ohms. When the current in this shunt is 8 amperes the current through the meter is

(A) 4 amperes
(B) 4 milliamperes
(C) 2 amperes
(D) 2 milliamperes.

55. In a bar-to-bar test applied to a D.C. armature you noticed that at a certain point around the commutator the voltage between the two adjacent commutator bars is almost equal to the applied test voltage. This is an indication that the armature coil between the two bars is

(A) shorted   (C) open
(B) grounded   (D) reversed.

56. A resistance of 20 ohms after being measured with an accurate bridge is found to be 20.05 ohms. It can be said that the error of this resistance is

   (A) 0.25%   (C) 1%
   (B) 0.5%   (D) 1.5%.

57. The viscosity of a lubricating oil is

   (A) the degree of fluidity of the oil
   (B) the lowest temperature at which the vapors given off will ignite
   (C) the amount of uncombined acid contained in the oil
   (D) the quantity of oily vapor an oil will give off at the temperature of a bearing.

58. The instrument which is commonly used to measure the specific gravity of a lubricating oil is a

   (A) calorimeter
   (B) hydrometer
   (C) barometer
   (D) viscosimeter.

59. When drilling holes perpendicular to the longitudinal axis of a round shaft it is necessary to use

   (A) V blocks
   (B) S blocks
   (C) a magnetic chuck
   (D) a universal chuck.

60. A frozen bearing is best removed from a shaft by first

   (A) heating both the shaft and the bearing
   (B) chilling both the shaft and the bearing
   (C) heating the bearing while keeping the shaft cool
   (D) chilling the bearing while keeping the shaft hot.

61. A valve that allows water to flow in one direction only is called a

   (A) globe valve
   (B) gate valve
   (C) check valve
   (D) needle valve.

62. Lubricants are oxidized when exposed to

   (A) water
   (B) air
   (C) acetylene gas
   (D) hydrogen.

63. When two identical centrifugal pumps are operated in series

   (A) the head at which they operate together at a given capacity is the same as that for a single pump
   (B) the capacity at which they operate at a given head is the same as that for a single pump
   (C) the efficiency at which they operate together at a given capacity is double that for a single pump
   (D) the head at which they operate together at a given capacity is double that for a single pump.

64. The velocity of discharge through a hole in the bottom of a tank containing 25 ft. of water is

   (A) 20 ft. per sec.
   (B) 40 ft. per sec.
   (C) 60 ft. per sec.
   (D) 80 ft. per sec.

65. The discharge of water through a certain pipe is 12,640 gal/min. If the velocity of flow is 133.7 ft./min., then the diameter of the pipe in feet is approximately (Note: 1 cubic ft. = 7.48 gals.)

(A) 4   (C) 2
(B) 3   (D) 1.

66. In water carrying capacity the number of 5" pipes equivalent to one 10" pipe is

(A) 1.5   (C) 3
(B) 2   (D) 4.

67. Even under perfect conditions, water cannot be lifted by suction more than

(A) 11.3 ft.   (C) 33.9 ft.
(B) 22.6 ft.   (D) 67.8 ft.

68. The temperature coefficient of linear expansion is

(A) the amount of expansion a 5-ft. rod undergoes during a temperature change of 5°
(B) the increase in length of a unit length during a 1° temperature rise
(C) the actual decrease in length of a 5-ft. rod of metal, divided by 5, divided by the decrease in temperature
(D) the amount of expansion of a 1 ft. rod during a heat change of 1 btu.

69. A five-part "block and fall" is being used to lift a short length of pipe which weighs 400 lbs. The theoretical pull in lbs. which one man would have to exert in order to lift the pipe is most nearly

(A) 60
(B) 80
(C) 100
(D) 120.

Venturi Tube

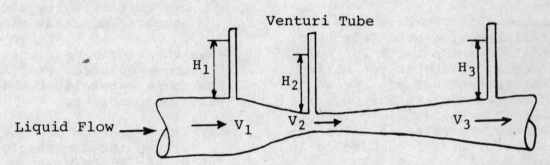

Questions 70 and 71 relate to the sketch of the venturi tube shown above.

70. With reference to the above diagram the height of the liquid in H_2 is

(A) greater than H_1
(B) less than H_3
(C) greater than H_3
(D) equal to H_1.

71. With reference to the above diagram the velocity at point V_2 is

(A) greater than V_1
(B) less than V_1
(C) less than V_3
(D) equal to V_3.

72. Assume that you are the Stationary Engineer (Electric) in charge of a watch. A new worker comes to you with a suggestion to change the procedure in doing some routine maintenance work. It is best that you should

(A) discuss the suggestion with the worker and determine its value
(B) tell him that you use only standard procedures and cannot consider his suggestion
(C) refuse to consider the suggestion because he is a new worker
(D) refer the matter immediately to your superior.

73. Assume that you are the

Stationary Engineer (Electric) in charge of a watch and that you have allowed one of your workers a privilege because it makes no difference in the running of this watch if one worker has this privilege. Serious complications arise later because several other workers want the same privilege. Your best action would be to

(A) withdraw the privilege originally granted
(B) grant the privilege to all the workers, thus avoiding favoritism
(C) tell the workers that you are the boss and show them why you are justified in giving this privilege
(D) temporarily deprive the worker of the privilege.

74. Assume that you have been the Stationary Engineer (Electric) in charge of a watch for several weeks when you notice that the workers are beginning to show a dislike for you, and that this is adversely affecting their morale and efficiency. Your best procedure is to

(A) loosen up the discipline of the watch
(B) insist on better discipline
(C) take stock of yourself to determine if you are to blame
(D) pay no attention because bosses are bound to be disliked no matter what they do.

75. The main reason for not using water to extinguish fires on or near high voltage electrical equipment is that water

(A) may ruin the insulation of the equipment
(B) may conduct electric currents and cause a shock hazard

(C) may increase the resistance of contacts by corroding them
(D) will form a dense cloud of steam and hinder firefighting efforts.

76. You start artificial respiration on a victim of electric shock, and shortly afterward another employee arrives at the scene of the accident and prepares to relieve you. In this case,

(A) you should skip one cycle to make the changeover
(B) the change should be made while your hands are on the victim applying pressure
(C) the change should be made while your hands are off the victim's body
(D) both operators should apply pressure together for two cycles before the second operator proceeds alone.

77. When the victim mentioned in the above question starts breathing naturally but is still unconscious,

(A) resuscitation must be continued ten minutes longer
(B) he should be given a stimulating drink immediately
(C) he should be propped up in a comfortable position
(D) he should be watched carefully in case breathing stops.

78. Assume that you are the Stationary Engineer (Electric) in charge of a watch, and that there is an unpleasant routine job to be done. The best procedure to follow in getting this job done is

(A) do the job yourself
(B) rotate the job among all the workers on your watch
(C) assign this job to one worker and give him special privileges as a compensation

(D) assign this job to a junior worker until he gains seniority.

79. Assume that you are the Stationary Engineer (Electric) in charge of a watch, and among your workers are several of your close friends. Some of the rest of your workers falsely accuse you of favoritism. In this situation your best course is to

(A) report the accusations to your supervisor denouncing the accusers
(B) ask the accusers to give specific details so that you can disprove them
(C) give your friends all the "dirty" jobs, thus disproving your favortism
(D) tell your accusers that you are the boss and you have faith in the ability of your friends.

80. You, as the Stationary Engineer (Electric) in charge of a watch, go on a tour of inspection with two of your workers, and find that another of your workers who was working alone has fallen off a high scaffold, and is lying unconscious on the ground. Of the following, the best and safest procedure is to

(A) have your two workers carry him to a warm office while you go to summon a doctor
(B) have one of your workers help you to prop him up in a comfortable position and give him a drink of water while the other worker goes to summon a doctor
(C) start artificial respiration immediately after administering a stimulant
(D) cover the worker with a blanket, send one worker to call the doctor and the other to get some smelling salts.

Answer Key

1.	C	17.	B	33.	C	49.	C	65.	A
2.	C	18.	D	34.	B	50.	C	66.	D
3.	D	19.	B	35.	A	51.	D	67.	C
4.	C	20.	B	36.	B	52.	B	68.	B
5.	C	21.	B	37.	D	53.	C	69.	B
6.	B	22.	A	38.	B	54.	B	70.	B
7.	B	23.	B	39.	C	55.	C	71.	A
8.	B	24.	C	40.	C	56.	A	72.	A
9.	D	25.	C	41.	B	57.	A	73.	A
10.	B	26.	C	42.	A	58.	B	74.	C
11.	B	27.	B	43.	D	59.	A	75.	B
12.	C	28.	B	44.	C	60.	C	76.	C
13.	A	29.	B	45.	B	61.	C	77.	D
14.	D	30.	C	46.	C	62.	B	78.	D
15.	B	31.	D	47.	B	63.	D	79.	B
16.	A	32.	B	48.	C	64.	B	80.	D

SAMPLE PRACTICE
EXAMINATION 12

STATIONARY ENGINEER (ELECTRIC)

DIRECTIONS: Each question has four suggested answers, lettered A, B, C, and D. Decide which one is the best answer, locate the question number on the sample answer sheet, and with a soft pencil darken the area that corresponds to the answer you have selected. All sample answer sheets follow page 22.

The time allowed for the entire examination is 3½ hours.

1. The direction of rotation of a D.C. shunt motor can be reversed by

   (A) reversing the line leads
   (B) reversing both the armature and field current
   (C) reversing the field or armature current
   (D) reversing the current in one pole winding.

2. The insulation resistance of the stator winding of an induction motor is most commonly measured or tested with a(an)

   (A) strobe   (C) megger
   (B) ammeter   (D) S-meter.

3. Assume that three ohm resistances are connected in delta across a 208-volt, 3-phase circuit. The line current in amperes will be most nearly

   (A) 30   (C) 17.32
   (B) 20.4   (D) 8.66.

4. Assume that three 12 ohm resistances are connected in wye across a 208-volt, 3-phase circuit. The power, in watts, dissipated in this resistance load will be most nearly

   (A) 4200   (C) 1200
   (B) 3600   (D) 900.

5. The one of the following knots which is most commonly used to shorten a rope without cutting it is the

   (A) clove hitch
   (B) diamond knot
   (C) sheepshank
   (D) square knot.

6. Assume that it is required to pump 40 mgd of water against a 65 ft. head. If the pump efficiency is 65%, the B.h.p. of this pump is most nearly

   (A) 920   (C) 460
   (B) 700   (D) 176.

7. Assume that a pump had to be shut down temporarily due to

132

trouble which was first reported by an oiler. The one of the following entries in the log concerning this occurrence which is <u>least</u> important is

(A) the time of the shutdown
(B) the period of time the pump was out of service
(C) the cause of the trouble
(D) the time the oiler came on shift.

8. At sea level the theoretical maximum distance in feet that water can be lifted by suction only is most nearly

(A) 12.00   (C) 33.57
(B) 14.70   (D) 72.0.

9. For good performance, while a lubricating oil is in use its neutralization number should

(A) keep rising
(B) remain about the same
(C) be greater than 0.1
(D) be greater than 2.0.

10. Cast iron castings that need repairing are usually repaired by the process known as

(A) electric arc welding
(B) electro-forming
(C) brazing
(D) resistance welding.

11. The term S.A.E. stands for

(A) Standard Auto Engines
(B) Standard Air Engines
(C) Society of Automotive Engineers
(D) Society of Aviation Engineers.

12. The parts of a large sewage pump that would most likely need repairs after the least number of hours of operation are the

(A) pump casings
(B) impellers
(C) wearing rings
(D) outboard bearings.

13. Assume that the power in a balanced 3-phase load is measured by the two wattmeter method and is read by means of two wattmeters, namely, W_1 and W_2. If the power factor of the load is .5 leading

(A) W_1 will read positive and W_2 will read negative
(B) W_1 will read negative and W_2 will read positive
(C) both W_1 and W_2 will read negative
(D) W_1 will read positive and W_2 will read zero.

14. The current in amperes of a 220-volt 5-hp., D.C. motor having an efficiency of 90% is most nearly

(A) 18.8   (C) 14.3
(B) 17   (D) 20.5.

15. A shunt generator having an armature current of 50 amperes, an armature resistance of .05 ohms and a generated e.m.f. of 222.5 volts will most likely have a terminal voltage of

(A) 172.5 volts   (C) 222.5 volts
(B) 220.0 volts   (D) 225 volts.

16. Assume that a 4-pole, 220-volt D.C. motor has a back e.m.f. of 215 volts and 4 armature paths between terminals. If the field flux per pole is suddenly decreased to one half of its former value, the motor speed in rpm compared to its original speed will be most likely

(A) decreased to about one quarter
(B) decreased to about one half
(C) doubled
(D) increased by one quarter.

17. The frequency of the voltage generated in a synchronous machine having 8 poles and running at 720 rpm is most nearly

(A) 120
(B) 72
(C) 60
(D) 48.

18. Assume that a synchronous converter has two slip rings and a direct current voltage of 313 volts between the brushes. The effective alternating voltage between slip volts is most nearly

(A) 220 volts
(B) 278 volts
(C) 330 volts
(D) 440 volts.

19. A newly appointed Stationary Engineer (Electric) attempted to make an emergency repair on a D.C. motor that had an open armature coil (lap-wound). He completely cut this coil in two, and disconnected it from both commutator bars. He then ran an insulated jumper large enough to safely carry the current between the two bars. This attempted emergency repair will

(A) result in an inoperative motor
(B) not significantly affect the normal running of the motor
(C) cause the motor to emit vicious purplish sparks at the commutator while running
(D) cause the motor to overheat excessively while running.

20. The purpose of full wave rectifiers is to

(A) produce A.C. current which contains some D.C.
(B) change D.C. current to A.C.
(C) produce D.C. current having an A.C. ripple of twice the input frequency
(D) produce only A.C. current

having twice the input frequency.

21. The temporary production of a substitute for a two-phase current so as to obtain a makeshift rotating field in starting a single-phase motor is called

(A) phase splitting
(B) pole pitch
(C) phase transformation
(D) pole splitting.

22. In a fully charged lead acid storage battery the active material in the positive plates is

(A) sponge lead
(B) lead carbonate
(C) lead acetate
(D) lead peroxide.

23. A heat exchanger commonly located between the low pressure and high pressure cylinders of an air compressor is used to

(A) lower the temperature of the compressor air
(B) increase the relative humidity of the compressor air
(C) decrease the relative humidity of the compressor air
(D) raise the temperature of the compressor air.

24. The one of the following instruments which is used for the determination of the velocity of air in ducts is the

(A) psychrometer
(B) pitot tube
(C) "U" gauge
(D) spherometer.

25. A high tension breaker (4160 volts) should be equipped with a mechanical interlock that will prevent the breaker from

being raised or advanced into, and it should be lowered or withdrawn from the operating position unless

(A) it is open
(B) it is closed
(C) the full load is connected
(D) a light load is connected.

26. For the operation of a high tension breaker (4160 volts) the suitable control voltage for best performance usually is

(A) 600 volts A.C.
(B) 600 volts D.C.
(C) 208 to 440 volts A.C.
(D) 70 to 140 volts D.C.

27. The equipment on which you would be most likely to find an unloader is

(A) a centrifugal water pump
(B) an air compressor
(C) a vacuum pump
(D) a steam engine.

28. The term Saybolt refers to a measure of

(A) specific gravity
(B) boiling point
(C) hardness
(C) viscosity.

29. Assume that a centrifugal fan running at 750 rpm delivers 20,000 c.f.m. at a static pressure of one inch. If this fan is required to deliver 28,000 c.f.m. at the same static pressure, it should be run at a rpm speed of most nearly

(A) 1500   (C) 1150
(B) 1250   (D) 1050.

30. The horsepower of a fan varies as the

(A) cube of the fan speed
(B) square of the fan speed
(C) square root of the fan speed
(D) cube root of the fan speed.

31. The gearing for transmitting power between two shafts at right angles to each other consists of two essential parts:

(A) two worm wheels
(B) a worm and bevel gear
(C) square root of the fan speed
(D) two bevel gears.

32. If a transmission main drive gear having 30 teeth rotates at 400 rpm, and drives a counter shaft drive gear at 300 rpm, the total number of teeth on the countershaft drive gear will be

(A) 30
(B) 40
(C) 60
(D) 80.

33. The one of the following faults of a C.B. main contact which is not a cause of over-heating of air circuit breakers is

(A) excessive pressure
(B) insufficient area in contact
(C) oxidized contacts
(D) dirty contacts.

34. The main reason that larger size electrical cables (such as #0000) are always stranded rather than solid is that they

(A) are more flexible
(B) are stronger
(C) have a higher conductivity
(D) have a higher specific resistance.

Questions 35 to 37 refer to the diagram of the auto transformer and data below:

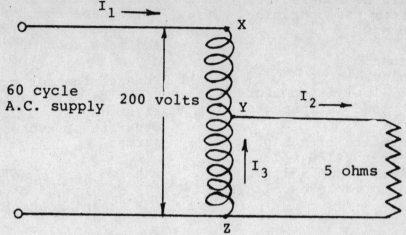

Data: An auto transformer whose primary is XZ is connected across a 200-volt A.C. supply as shown. The load of 5 ohms is connected across Y and Z. (Assume that point Y is the midpoint of the winding.)

35. The current I_1 in amperes is approximately equal to

 (A) 5   (C) 15
 (B) 10   (D) 20.

36. The current I_2 in amperes is approximately equal to

 (A) 5   (C) 15
 (B) 15   (D) 20.

37. The current I_3 in amperes is approximately equal to

 (A) 5   (C) 15
 (B) 10   (D) 20.

38. Assume that you see one of your oilers tumble down a long flight of concrete steps, and fall heavily on the lower landing. You rush to him, and find that he is unconscious but breathing. Of the following the best course of action for you to take is

 (A) have two of your men carry him to the office and summon a doctor
 (B) do not move him but cover him with a blanket and call a doctor
 (C) prop him upright and let

him inhale spirits of ammonia and call a doctor.
 (D) prepare a bed of blankets and have two of your men lift him on it, then summon a doctor.

39. It is sometimes desirable to have a control that will cause a D.C. motor to come to a standstill quickly instead of coasting to a standstill after the stop button is pressed. This result is most commonly obtained by means of an action called

 (A) counter e.m.f. method
 (B) armature reaction
 (C) diverting
 (D) dynamic braking.

40. In an electric circuit a high spot-temperature is most commonly due to

 (A) an open circuit
 (B) a defective connection
 (C) intermittent use of circuit
 (D) excessive distribution voltage.

41. The main reason for periodic inspections and testing of equipment in an electrically

powered plant is to

(A) keep the men busy at all times
(B) familiarize the men with the equipment
(C) train the men to be ready in an emergency
(D) discover minor faults before they have a chance to become serious.

42. Assume that an employee calls up to give advance notice of his intentions to be absent the following day. The most important information that he should give is

(A) the exact time of calling
(B) the balance of his sick leave time
(C) the reason for his absence
(D) name of attending doctor.

43. The main reason why a Stationary Engineer (Electric) assigned to service equipment must be able to make proper adjustments and repairs quickly is that

(A) equipment always deteriorates rapidly unless readjusted immediately
(B) idle equipment will result in poor plant efficiency and work delays
(C) the ability to work rapidly is the result of extensive training and experience
(D) he will have more time for his other duties.

44. A 1300-volt, 3-phase system with a grounded neutral has a phase to ground voltage of approximately

(A) 440   (C) 690
(B) 600   (D) 750.

45. A 220-volt 40 hp induction motor is given an insulation resistance test. The normal value of the insulating resistance, in megohms, for this motor, is most nearly

(A) 0.2   (C) 0.05
(B) 0.4   (D) 0.95.

46. To increase the range of an A.C. ammeter the one of the following which is most commonly used is a(an)

(A) current transformer
(B) inductance
(C) condenser
(D) straight copper bar.

47. When batteries are being charged, they should not be exposed to open flames and sparks because of the flammability of

(A) hydrogen
(B) oxygen
(C) sulphurous gas
(D) fuming sulphuric acid.

48. Assume that you and your supervisor are on an inspection tour of the outdoor equipment of the plant, and that a co-worker suddenly falls unconscious on the pavement. If on close observation you find that the victim is not breathing, the first of the following things to do is

(A) move the victim indoors
(B) notify his family
(C) administer first aid to restore breathing
(D) nothing, but summon a doctor.

49. Assume that one of your men, who has always been efficient, industrious, and conscientious suddenly becomes lax in his work, makes numerous mistakes, and shuns responsibilities. The cause of such a change

(A) is usually that the man is responding to a minor change in the job situation

(B) is usually apparent to the stationary engineer in charge and fellow workers

(C) may be quickly found by a close study of reports and personnel records

(D) may have no direct relationship to any change in the job situation.

Questions 50 to 52 refer to the diagram of a 3-phase transformer and data given below:

PRIMARY 1320 VOLTS

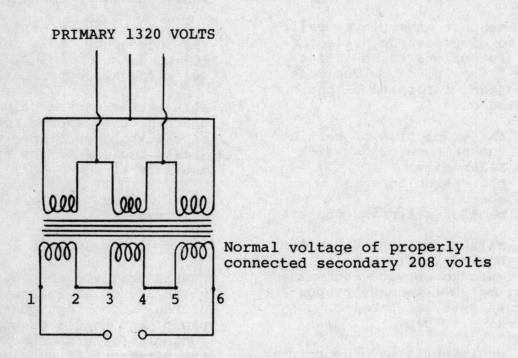

Normal voltage of properly connected secondary 208 volts

Data: The above transformer is to be connected delta-delta, with primary connections completed as shown. Assume that the connections of the secondary of the transformer bank are not completed and it is found that coil (1 - 2) is reversed. Under this condition:

50. The voltage between points 6 and 3 will be most nearly

(A) 208   (C) 416
(B) 360   (D) 520.

51. The voltage between points 1 and 6 will be most nearly

(A) 208   (C) 416
(B) 360   (D) 520.

52. The voltage between points 1 and 4 will be most nearly

(A) 208   (C) 416
(B) 360   (D) 520.

53. The one of the following types of valves which is generally used where extremely close regulation of flow is needed is the

(A) gate valve
(B) globe valve
(C) needle valve
(D) blow-off valve.

54. Lubricating oils that are of mineral origin are refined from

    (A) lard-beef products
    (B) cotton seed products
    (C) crude petroleum products
    (D) lime soap products.

Questions 55 to 57 refer to the diagram below:

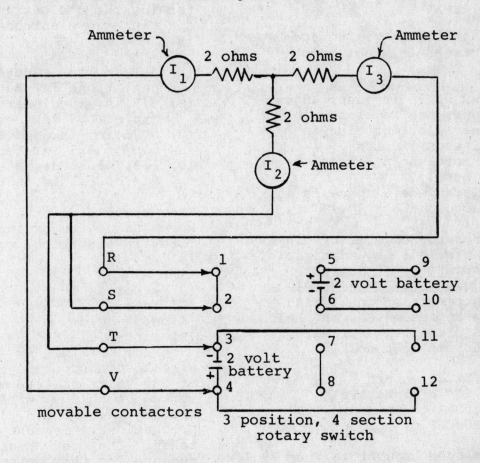

55. When switch movable contactors R, S, T, and V are in position 1, 2, 3, and 4 as shown, the current I_1 in amperes is most nearly

    (A) 2   (C) 1/3
    (B) 2/3   (D) 1/6.

56. When switch movable contactors R, S, T, and V are in position 5, 6, 7, and 8, the current I_2 in amperes is most nearly

    (A) 2   (C) 1/3
    (B) 2/3   (D) 1/6.

57. When switch movable contactors R, S, T, and V are in position 9, 10, 11, and 12, the current in amperes registered by ammeter I_3 is most nearly

    (A) 3   (C) 2/3
    (B) 2   (D) 1/3.

58. Light-bodied lubricating oils are most commonly used for

    (A) light loads at high speeds
    (B) heavy bearing pressure
    (C) heavy loads at slow speeds
    (D) chain drives and gears.

59. The one of the following lubricants which is least likely
to be attacked by acids is

    (A) cotton seed oil
    (B) castor oil
    (C) rape seed oil
    (D) graphite.

60. In general, non-rising stem
gate valves are best adapted
for

    (A) use where frequent adjust-
    ments are necessary
    (B) installations carrying
    viscous liquids
    (C) throttling or close
    control
    (D) places where space is a
    factor.

61. The presence of moisture in
insulating oil is undesirable.
The percentage of moisture
which will reduce the dielectric strength of insulating
oil to approximately one half
of its dielectric strength
when dry is most nearly

    (A) 0.5% of moisture
    (B) 0.05% of moisture
    (C) 0.005% of moisture
    (D) 0.0005% of moisture.

62. It has been brought to your
attention that one of the men
under your supervision is com-
plaining to fellow co-workers
that another man has received
an easy assignment through
his "connections." In this
situation it is best to

    (A) privately inform the man
    who is complaining of the
    truth regarding the assign-
    ment
    (B) in the presence of others,
    demand absolute proof from
    the man who is complaining
    (C) ignore the matter since it
    is not your job to inter-
    fere in disagreements be-
    tween the men

    (D) tell the complaining man
    to apply for a desirable
    assignment also.

63. In the standard method of
testing electrical insulating
oils the test cup used to determine the dielectric strength
contains two electrodes. Each
is

    (A) 0.1 in. in diameter with a
    gap of one in. between them
    (B) 0.5 in. in diameter with a
    gap of 0.3 in. between them
    (C) 0.75 in. in diameter with a
    gap of 0.3 in. between them
    (D) 1.0 in. in diameter with a
    gap of 0.1 in. between them.

64. The proper fire extinguishing
agent to use to extinguish
fires in electrical equipment
is

    (A) water
    (B) foam
    (C) soda-acid
    (D) carbon dioxide.

65. Circuit conductors operating
at 600 volts or less may be
worked upon live, without
opening the circuit, if certain
precautionary measures are
taken. The best precautionary
measure for this work is

    (A) bare or exposed places on
    one conductor must be
    taped after another con-
    ductor is first exposed
    (B) adjacent live or grounded
    conductor must be covered
    with a conducting material
    (C) bare or exposed places on
    one conductor must be
    taped before another con-
    ductor is exposed
    (D) adjacent live or grounded
    conductors must be securely
    bonded to ground.

66. In order to properly distribute
the load (in proportion to
their rated capacities) between

two alternators that are operating in parallel it is necessary to

(A) overexcite the smaller alternator and underexcite the larger one
(B) adjust the governor on the prime mover
(C) underexcite the larger alternator but use normal excitation on the smaller one
(D) underexcite the smaller alternator and overexcite the larger one.

67. If a large amount of flame is visible from a small pile of burning material it is likely that the material must contain a substance that

(A) contains a large amount of inorganic material
(B) produces during the burning process a large amount of pure carbon
(C) produces during the burning process a large amount of combustible gases or vapors
(D) is composed almost entirely of pure carbon.

68. If the velocity of water flow in a pipe is doubled, assuming other factors are constant, the loss of head due to friction will be

(A) decreased 1/2 times
(B) decreased 1/4 times
(C) increased 4 times
(D) the same.

69. Reprimanding a subordinate for inefficiency in the presence of fellow co-workers is apt to

(A) cause the subordinate to resign
(B) arouse the subordinate's resentment
(C) improve the performance of all present
(D) cause the subordinate to improve.

70. Assume that certain work assignments are not liked by any of your subordinates. Because this work has to be done, you as the Stationary Engineer (Electric) should try as much as possible to

(A) assign this work as punishment details
(B) rotate the work assignments among subordinates
(C) assign this work to the best-natured worker
(D) assign this work to the junior workers.

71. A Stationary Engineer (Electric) when discussing new department regulations with subordinates commented, "We should be conscious of the fact that our interests are mutual, and that by all of us in unison putting our shoulder to the wheel and working together, we can achieve our common objective." This approach is

(A) good, because this attitude will promote cooperation
(B) poor, because this approach will invite excessive criticism
(C) good, because it will promote good fellowship
(D) poor, because this will invite too much familiarity.

72. In the inspection of relays the type of trouble generally encountered often depends on the type of relay. The one of the following which is not a trouble encountered with an induction-type relay is

(A) friction between disc and magnet
(B) dust on disc
(C) foreign matter in the gear train
(D) punctured bellows.

73. With reference to diesel engines the one of the following which is not a method of scavenging the cylinder is

(A) crankcase scavenging
(B) integral scavenging
(C) under-piston scavenging
(D) vane scavenging.

74. Direct current motors for best performance should have their brushes set on the commutator

(A) at the neutral point (under load)
(B) at the point of maximum armature reaction
(C) radially at an angle of 90° (leading)
(D) radially at an angle of 80° (leading).

75. The proper order of events that take place in a 4-stroke cycle diesel engine is

(A) air intake, power expansion, compression, and exhaust
(B) air intake, compression, power expansion, and exhaust
(C) power expansion, air intake, compression, and exhaust
(D) compression, air intake, power expansion, and exhaust.

76. The compression ratio of a diesel engine that has no starting ignition device is generally in the range of

(A) 11 to 20   (C) 6 to 8
(B) 8 to 10   (D) 4 to 6.

77. The base in a lubricating grease denotes the

(A) type of soap that is used in its manufacture

(B) consistency and the texture of the grease
(C) dropping or melting point of the grease
(D) carbon-residue content of the grease.

78. Of the following sets of pipes the one having a total combined area exactly equal to the area of a 12" diameter pipe is

(A) two 6" pipes
(B) two 8" pipes
(C) one 8" pipe and two 6" pipes
(D) four 6" pipes.

79. Assume that a single-phase load takes EI x .8 watts, where E is the line voltage, I the line current, and .8 the power factor. The rating in volt-amperes of the synchronous condenser needed to raise the power factor to unity is most nearly

(A) EI x .6
(B) EI x .8
(C) EI x .9
(D) EI x 1.

80. If rubber gloves commonly used on high tension work are found on test to have pin holes, they

(A) may be used on low voltage
(B) should be discarded
(C) should be patched with rubber tape
(D) may be used only in dry places.

Answer Key

1.	C	17.	D	33.	A	49.	D	65.	C
2.	C	18.	A	34.	A	50.	A	66.	B
3.	A	19.	B	35.	B	51.	C	67.	C
4.	B	20.	C	36.	D	52.	B	68.	C
5.	C	21.	A	37.	B	53.	C	69.	B
6.	B	22.	D	38.	B	54.	C	70.	B
7.	D	23.	A	39.	D	55.	B	71.	A
8.	C	24.	B	40.	B	56.	C	72.	D
9.	B	25.	A	41.	D	57.	D	73.	D
10.	C	26.	D	42.	C	58.	A	74.	A
11.	C	27.	B	43.	B	59.	D	75.	B
12.	C	28.	D	44.	D	60.	D	76.	A
13.	D	29.	D	45.	A	61.	C	77.	A
14.	A	30.	A	46.	A	62.	A	78.	D
15.	B	31.	D	47.	A	63.	D	79.	A
16.	C	32.	B	48.	C	64.	D	80.	B

PART THREE

Background and Study Material

POWER PLANT OPERATION

Part 1: Boilers

Maximum Capacity: What is meant by the maximum capacity of a boiler?

Maximum capacity is the greatest amount of power which the boiler can develop when it is burning all the fuel it can, with the highest draft available. This may be rated either as horsepower, which is determined either by dividing the total actual evaporation per hour by 30, or by a percent of rated boiler capacity, which is found by dividing the actual horsepower developed by the nominal rated horsepower.

Horsepower: How do you calculate the horsepower of a boiler when the total pounds of steam generated is known?

Boiler horsepower is defined as the equivalent evaporation of 34.5 lbs. of water from 212°F. to steam at 212°. It is necessary to first determine the equivalent evaporation of the boiler under the conditions of evaporation.

Assuming the boiler to generate 1000 lbs. of steam per hour having a quality of 98% and operating at a pressure of 100 lb. gauge with feed water at 150°F., we first calculate the amount of heat absorbed by the water in the boiler.

One thousand lbs. of water are raised from 150° to 337.9°, the boiling temperature at 100 lb. gauge or 114.7 lb. absolute. Hence we have (337.9 -- 150) x 100 equals 187,900 Btu to raise the water to boiling temperature.

Latent heat of one lb. of steam according to Goodenough's Steam Tables is 882.3 Btu. Since the steam contains 2% of moisture, only 98% of the 1000 lbs. delivered is steam, so 980 x 882.3 = 864,650 Btu are used to convert the water into steam. The total heat then is 187,900 plus 864,650 = 1,052,550 Btu absorbed by the boiler. The latent heat of steam at 212° is 970.3 Btu, as to psig. Therefore, dividing 1,052,550 by 970.3 gives 1,803.2 as the equivalent evaporation from and at 212° and 1,083 divided by 34.5 gives 31.4, which is the horsepower of the boiler under these conditions.

Horsepower and Pressure: Of two boilers carrying the same load in boiler horsepower, one at 75 lb. gauge pressure, the other at 125 lb., which will evaporate the greater amount of water?

With feed temperature the same, the evaporation will be greater for the lower pressure. A boiler horsepower is 34.5 of water evaporated from and at 212°F. The factor of evaporation that is the ratio of equivalent evaporation from and at 212° to actual evaporation under operating conditions is: F = (H⁻ʰ) divided by 970.4, where H is the total heat in steam at boiler pressure and h is sensible heat in feedwater. Assuming feedwater at 200°, h is 168. For 125 lb. gauge, H is 1193.7. For 75 lb. gauge, H is 1186.5.

Then for the 125 lb. boiler, F is (1193.7 - 168) divided by 970.4 = 1.057, and for the 75 lb. boiler, F is (1186.5 - 168) divided by 970.4 or 1.048. Dividing 34.5 by the factor of evaporation gives for the 125 lb. boiler, 32.65 lb. actual evaporation per horsepower, and for the 75 lb. boiler, 32.86 lb. per horsepower.

(Questions and answers in this portion of the book have been compiled by the editorial staff of Power Plant Engineering Magazine and have been copyrighted by that publication. They appear with the permission of the magazine.)

Ordinarily the lower pressure boiler would have feedwater at lower temperature. If feed temperature for the 75 lb. boiler be taken the same amount lower as the steam temperature, it would be 170° and h would be 137.9. Then F equals (1186.5 - 137.9) divided by 970.4 or 1.081, and 34.5 divided by 1.086 equals 31.9 lb. actual evaporation per horsepower, which would be less than that for the 125 lb. boiler with feedwater at 200°.

BOILER SETTINGS AND PROPORTIONS

Boiler Suspension: What are the advantages of suspending a steam boiler by hangars?

A boiler suspended by hangars is free to expand and contract without affecting the brickwork of the setting. While a boiler properly supported on the brickwork with rollers (to take care of the motion) will work without cracking the setting, it is freer to move if supported by hangars. Also, any setting of the brickwork does not tend to throw the boiler out of level, and replacing the brickwork is more easily done.

Bridge Wall: How is the distance from the top of the bridge wall to the shell of a return tubular boiler determined?

It is somewhat arbitrary, the distance being greater if the boiler is to carry overload than where operating is not to exceed rating. An old rule of thumb gave the distance as 1/7 the grate area, flue area being 1/8, and the chimney area 1/9 grate area. Modern practice would give the distance from wall to shell as 1/5 to 1/4 the grate area, which allows for passing a greater volume of gas, and avoids the tendency to force gases against the shell.

CONSTRUCTION

Steam Outlet: At which end of a boiler should the main steam outlet be placed?

The main steam outlet is placed where it is most convenient. The essential condition is that the steam be drawn from the high and dry point in the steam drum.

Reheat: What is a reheat boiler?

Reheating of steam after partial expansion is done to evaporate moisture, and to add superheat so that steam will be dry when it reaches exhaust. The boiler for this purpose also generates steam in the lower portion, the reheat being really a large additional superheater that takes the place of steam-making surface. In one common arrangement the reheat section is above the regular steam-making and superheating section. Steam from the boiler drum is sent through the high-pressure superheater to the steam mains. The reheat section has no connection to the rest of the boiler, but receives exhaust from the high-pressure turbine and delivers it to the low-pressure turbine supply main.

Steam Nozzle: What is the method for determining the correct size of a steam nozzle for a boiler?

The maximum steam that the boiler will be called upon to deliver should be passed by the nozzle at a given velocity, usually taken at 800 ft. per min. For a 500 hp. boiler to operate at 200% of rating, or 1,000 hp. and at 150 lb. pressure, feedwater 200°, the steam will have a volume of 2.76 cu. ft. per lb. Evaporation per hr. from and at 212° would be 1000 x 34.5 or 34,500 lbs., and the factor of evaporation would be 1.059. Hence actual delivery would be 34,500 divided by 1.059 or 32,600 lbs. per hr. Volume to be passed per hr. would be 32,600 x 2.76 or 90,000 cu. ft., or 1,500 cu. ft. a min. At a velocity of 8,000 ft. per min., the area needed would be 1,500 divided by 8,000, which equals 0.1875 sq. ft. or 27.0 sq. in. From a table of circular areas this would call for a diameter of nozzle of 5.86 in.

Feed Pipe Location: Where does the feed pipe enter a B&W boiler and how is it supported?

The feed pipe in a B&W boiler enters the boiler through the lower part of the front drum head, and is extended beyond the front tube header through the baffle plate at the head of the boiler; it is, therefore, partly supported by the latter. It is also supported by a stay bolt near the end that is attached to the feed pipe by a coupling, and extended upwards through the baffle plate to which

it is fastened by a nut.

Blowoff Line: On the blowoff lines of a water-tube boiler is a common cock next to the boiler, with a blowoff cock on its outside. Is this correct? Wouldn't it be safer and better to have a screw valve on the outside?

Proper blowoff connection requires the use of a cock nearer the boiler in the flowoff line, and a blowoff valve just beyond the cock (two valves in series are sometimes used).

Blowoff Tank: What size blowoff tank is needed for a battery of three 100 hp. boilers? Is brick or concrete suitable?

It is usual to provide, in a blowoff tank or basin, about 1/24 the volume of the boilers to be blown off into it simultaneously. In case of three boilers it is likely that all of them might not be blowing at the same time, or before the tank could be emptied.

The volume of a 100-hp. boiler would be about 320 cu. ft., or for three, 960 cu. ft. 1/24 of this would be 40 cu. ft., the capacity desirable for the basin. This is on the supposition that the basin will be emptied into a sewer or other drainage after the water has had a chance to cool and give up its steam.

There would be no objection to the use of concrete, provided a good coat of neat cement is put on the inside of the tank. This should be at least 1/2 inch thick. Brick would hardly be desirable, although it would be possible to build the tank of brick, laying it up with Portland cement mortar, and then to give this a coating on the inside of neat cement 1/2 inch thick.

Water Column Connection: How high above the top row of flues should the first gauge cock on a horizontal return be placed?

Each boiler should have at least one water gauge glass with connections not less than 1 in. pipe size. The lowest visible part of the water glass shall not be less than 3 in. above the top of tubes and 2 in. above the fusible plug. The water gauge glass should be equipped with a valved drain. The lowest permissible water level should be that at which there will be no danger of overheating any part of the boiler when it is operated with water not lower than that level.

Each boiler should have three or more gauge cocks located within the range of the visible length of the water glass, except when the boiler has two water glasses with independent connections to the boiler, and is located on the same horizontal line and not less than two feet apart.

Gauge Cocks: What is the object of placing the gauge cock outlets on an angle on the head of a steam boiler?

Try-cocks on a boiler are placed at an angle for two reasons--to direct the discharge away from the operator and also from the cock immediately below.

Nonreturn Valve? Where is a non-return stop valve placed on a boiler?

Nonreturn stop valves are placed on the top of the boiler. The A.S.M.E. Boiler Code has the following specification regarding installation of stop valves: "When two or more boilers, having manhole openings, are connected to a common steam main, two stop valves with an ample free-blow drain between them, shall be placed in the steam connection between boiler and the steam main. The discharge of this drain valve must be visible to the operator while manipulating the valve. The stop valves shall consist preferably of one automatic nonreturn valve (set next to the boiler) and a second valve of the outside screw-and-yoke type; or two valves of the outside screw-and-yoke type, may be used."

Automatic nonreturn valves stop a flow of steam from the main heater into the boiler in case of rupture of any part of the boiler. They also automatically cut the boiler in and out of service. When the pressure in the boiler becomes as great as that in the main header, the valve opens automatically.

Boiler Top Covering: What is a good and cheap covering for the tops of boilers?

A good, cheap and efficient covering for tops of boilers so that the heat losses from this source can be reduced to a minimum is: First cover the top with coarse ashes from 6 to 8 in. thick; then cover this with a cement consisting of four wheelbarrows of coarse ashes, one bag Portland cement, one bag of asbestos cement thoroughly mixed dry and wet. This makes a top sufficiently hard to walk on.

Staybolt Location: What are stay-bolts, and in what part of a steam boiler are they used?

Staybolts are, or should be, made of the best quality of wrought iron, their diameter being 3/4 in., 7/8 in., or 1 1/4 in., depending on the size of the boiler and the pressure it is to carry. They have a screw thread cut their entire length, and are screwed through both the inside and outside plates of fire box boilers at intervals of 4 to 4 1/2 in. from center to center, thus securely binding the plates together. The ends are allowed to project through the plates far enough to permit of their being riveted down on to the plates. In some cases they are made hollow, so that a broken bolt will be shown by leakage. To avoid threading the entire length the ends are sometimes upset.

Staybolt Size: How is the proper size of a staybolt determined?

To determine the proper size of a staybolt, first multiply the pressure in lbs. per sq. in. by the square of the pitch of the staybolts. This gives the total pressure which the staybolt has to carry.

Divide this total pressure by the allowable stress per sq. in. on the staybolt, and this gives the area of the bolt at the bottom of the thread. From this area find the size of staybolt permissible. In some states the allowable stress per sq. in. for staybolts 1 1/4 in. in diameter and under is 6,500 lb.; over 1 1/4 in. diameter, 7,000 lb.

Stay Bowing: Why do the thorough stays on a horizontal return tubular boiler bow up?

When installed, all through stays are intended to be perfectly straight, but due to contraction and expansion of other parts of the boiler, they may become bowed up. If the fire sheet bulges out due to heat, as it frequently does, the lower part of the front head is pulled inwardly, causing the through stays to be bowed away from the fire sheet.

Rivet Spacing: Why are small, closely spaced rivets used in a boiler joint instead of widely spaced large ones?

Two things must be accomplished in the boiler joint: To hold the joint tight so that it can be caulked to hold pressure, and to hold the parts together. Although the strength against shearing or tearing apart might be accomplished with the large rivets widely spaced, the edges of the sheet would buckle so that the pressure would force the sheets apart.

Rivet Length: How long should a rivet be before it is driven to form the proper size head?

The length of the rivet before it is driven depends, of course, upon the size of the rivet, the size of the hole and the thickness of the joint. Complete tables have been made up for the use of boiler manufacturers, and give the length of rivets for various size joints. Boiler rivets when driven must completely fill the hole in which they are put, and from a head that has the same strength as the tensile strength of the body of the rivet.

Dry Sheet: What part of a tubular boiler is known as the dry sheet?

In the slush front horizontal tubular boiler the smoke box at the front end of the boiler is over the fire doors, and the part of the sheet that extends beyond the front head is called the "dry sheet." This is not protected by water on the upper side, but has a firebrick lining on the under side to keep the heat of the furnace away from it.

SUPERHEATER

Heat Added: What is the percentage of heat added by a superheater to steam of 140 lb. gauge pressure, and of 97% quality, in order to obtain steam of 470°F. at that pressure?

For 97% saturated steam at 140 lb. gauge, which is equivalent to 155 lb. ads., the following heats exist:

Heat of liquid	361.00
Latent heat of evaporation (861 x 0.97)	835.17
Total heat in 1 lb. wet steam	1196.17

Total heat in 1 lb. superheated steam at 155 psi a and 470°   1257.3

Heat added to wet steam by superheater equals 1257.3 minus 1196.17, or 61.13. Then 61.13 divided by 1257.3 equals 0.0486, or the answer, 4.86%.

Superheaters: What would be the suitable installation of a super-heater in a return tubular boiler setting, when the distance from the back of the boiler to rear wall is 20 in.?

For this type of boiler the usual installation for a superheater is in the rear gas pass. Whether 20 in. is sufficient to avoid restriction of gas flow cannot be stated definitely, but probably more space, say 3 to 4 ft., would be better.

Superheater Safety Valves: Should the safety valve on a superheater be set for a higher or lower pressure than that on the boiler? Why?

Safety valves on a superheater are set to blow at a lower pressure than the setting of the boiler safety valves so that in case the flow of steam through the steam valve pipe ceases and the fire is not checked, the safety valve on the superheater will blow and set up a circulation to prevent overheating of the su-perheater. Where the boiler with a superheater has several safety valves, it is customary to set one of the boiler safety valves to blow just in advance of the superheater safety valve, so that with the boiler under normal operating con-ditions and delivering steam into the header, the boiler safety valve will blow and give warning to check the draft to prevent the superheater safety valve from blowing and pos-sibly drawing sediment into the header.

Part 2: Furnaces and Drafts

DRAFT

Draft Required: Is there a rule to get the draft over the grate that will give the proper amount of air for combustion?

There is no definite rule. The draft over the grate should be as low as will furnish air to give maximum CO_2 at as constant a rate as pos-sible. In practice it will vary from 0.05 to 0.25 in. of water. If greater, air leakage through the setting increases and efficiency decreases. If positive pressure is allowed above the grate, furnace walls are overheated.

AIR SUPPLY

Air Supply Needed: What air sup-ply is used for pulverized fuel? One type of burning supplies about 10% of the air for combustion, with the fuel through the burner at pressure from 5 to 10 in. of water. Another 25% comes in around the burner and the balance through openings in the furnace walls at about 1 in. pressure.

Bridge Wall Air: Can coal contain-ing 36% volatile combustible be burned smokelessly by passing cold air through the bridge wall from a point below the grate to the com-bustion space back of the bridge wall?

It is better to attain smokeless combustion by other means. Care-ful spread firing, admitting air over the fire and opening the stack dampers for a short time (2 to 5 min.) after firing, study of fire bed thickness, and use of a flue gas analyzer and a draft gauge will be more effective. Air admitted to the combustion cham-ber is too late to be useful un-less it comes in at a temperature of 400° to 500°F., and only a small quantity could be heated to such a temperature in passing through the bridge wall.

Actual heat loss from smoke is probably small, but with heavy firing smoke is often accompanied by large quantities of unburned hydrocarbon gases. This can be remedied by lighter and more fre-quent firing.

Preheating Air: What gain is realized by preheating air with flue gases?

In tests made at Colfax Station and reported by C.W.E. Clarke to the A.S.M.E., on a B&W cross drum boiler rated at 79.053 lbs. of steam an hour from and at 212°, with air preheated to 120° at 114% of rating and 140° for 200% of rating, results were as follows:

CO_2 increased 0.7% at low output and 1.4% at high output. Decrease of combustible in ash 6% and 12%. Decrease in draft loss of 20% and 30%. Temperature of gases higher up to the end of the third pass. Efficiency increased 5.7 and 7%.

FURNACE CONSTRUCTION

Radiation Loss: How is radiation loss through boiler furnace walls determined or estimated? What percentage of fuel fired would be wasted by uninsulated wall, say 9 in. firebrick and 8 in. common brick, operating at 150% of rating?

Various authorities give figures from 2 to 6% for this loss at rating. From the total of 1) the heat available from the coal, 2) the heat absorbed from the boiler, and 3) heat lost through evaporation of moisture in the coal and air, subtract the amount lost through steam formed from hydrogen in the coal, through flue gases, and through incomplete combustion as shown by carbon monoxide in gases and combustible in ashes. The remainder is loss due to unburned hydrogen and hydrocarbons, to radiation and other unaccounted losses, but there is no way of separating these.

Radiation increases with the increasing load, but radiation per horsepower decreases. It depends on the surface area, thickness, construction and material in the walls as well as arrangement of boilers in battery.

Furnace Temperature: What are maximum theoretical and practical temperatures in a boiler furnace, and what conditions affect these temperatures?

Conditions affecting furnace temperature are: 1) nature and quantity of the fuel fired, including the moisture present; 2) nature and quality of the products of combustion; 3) specific heat of the gases formed including excess of air; 4) radiation from the furnace walls.

Carbon combines with oxygen, 12 parts by weight of carbon to 32 parts oxygen, resulting in 44 parts carbon dioxide, or a lb. of carbon uses 2.67 lbs. oxygen and forms 3.67 lbs. CO_2. Air is 77% nitrogen and 23% oxygen, so to furnish 2.67 lbs. oxygen requires 11.57 lbs. air, of which 77% or 8.9 lbs. is nitrogen for each lb. of carbon burned.

Neglecting radiation, all heat liberated will go into the products of combustion. The specific heat of CO_2 is 0.216, and that of nitrogen is 0.244, so for each degree rise in temperature, the CO_2 will require 3.67 x 0.216 or 0.792 Btu, and the nitrogen will require 8.9 x 0.244 or 2.17 Btu, for a total of 2.962 Btu (CO_2 + N_2 = 2.962 Btu).

One lb. of carbon burned to CO_2 gives up 14,600 Btu so that the temperature rise will be 14,600 divided by 2.962 or 4929°F. above the temperature of the air. For 70° atmospheric, the maximum theoretical temperature would be 4929 plus 70, or 4,999°F.

Excess air must always be used. If 100% excess is supplied, which is not uncommon at specific heat of 0.24 for the air, it will require

11.57 x 0.24 or 2.78 Btu. to heat this excess one degree. To raise the total flue gases for 1 lb. carbon one degree would require $CO_2 + N_2 + air = 5.742$ Btu. Temperature rise would be 14,600 divided by 5.742, or 2540°F., and maximum temperature 2543 plus 70, or 2,613°F. However, coal, not carbon, is the fuel and this brings into the problem hydrogen and water both as to combustion heat and temperature rise. An equation which will give approximate results is:

T equals (616 x C plus 2220 x H - 327 x O - 44 x W) divided by (L.C.F. plus 0.02 x 2 plus 0.18 x H).

T is final temperature; CHO and W are percentages of carbon, hydrogen, oxygen and moisture in the fuel, and F is the lbs. of dry gases per lb. of fuel.

The effect of radiation is difficult to determine or to estimate. Also, the rate of combustion has an important effect on the drop in temperature due to radiation.

Mortar For Firebrick: What is a good mortar for laying up firebrick in a furnace?

A mixture of 1/2 fire clay and 1/2 fire brick, crushed to contain some dust, some fine grain and the remainder the size of large wheat is good. Mix with water to a thin mortar. For 1000 firebricks 1/2 ton of clay and 1/2 ton crushed firebrick will make the right quantity of mortar. Fusion point of the mortar will be about 3180°F.

STOKERS AND GRATES

Underfeed Stoker Action: An 850-hp. Sterling boiler has a multiple retort underfeed stoker burning Illinois coal, driven by a variable speed motor. Air pressure in the main wind box is 3 in. of water; draft over the fire is regulated by the damper to zero. CO_2 in the last pass is 14%. When load comes on, steam pressure drops, the stoker speeds up, the damper opens several notches and CO_2 drops to 9 or 10%. Main box air pressure does not change. What changes take place in the furnace under these conditions? Are combustion rate and steaming capacity increased? Why does CO_2 content decrease?

As to CO_2 changes, when the stoker is speeded up and the damper opened, more excess air enters the furnace. More rapid fuel feed will at first cause large openings in the fuel bed, which will lower wind box pressure and increase draft over the fire. If the increased speed is maintained, fuel bed openings will be closed after a time and unless more pressure is supplied in the wind box, the air supply will be deficient. The combustion rate will be lower, and draft over the fire will increase. This can be overcome somewhat by grate bar movement to break up the fuel bed, but this will allow pressure to build up over the fire, and the damper should be opened to maintain the same draft as before. More excess air is likely to be drawn in, which will lower the CO_2 and, even if the combustion rate is increased, flue gas losses may increase more than the heat absorbed by the boiler, so that steam pressure will drop.

Better operation, by either manual or automatic control, would be obtained (when load increases and pressure drops) by speeding up both stoker and fan so as to increase feed and wind box pressure. Then open the damper just enough to maintain the same draft as before over the fire. Driving the fan and stoker from the same engine, motor or turbine whose speed is controlled by steam pressure will vary feed and pressure together. Damper position can best be controlled by the draft over the fire by means of a pilot valve.

Grate Surface: What is the average allowance of grate surface to the horsepower?

In Massachusetts, for example, regulations provide that for boilers carrying 25 lb. pressure per sq. in. or less, 2/3 sq. ft. of grate surface shall be allowed per hp.; for boilers carrying over 25 lb., 1/3 sq. ft. of grate surface per hp. This difference is based on the fact that the low pressure, or heating boiler, usually has a low draft, and is run at a very low rate of combustion per sq. ft. of grate per hr.

Size of Grate: What determines the grate surface?

The length and width of a grate depend entirely on the size of the boiler and its construction. The length should not be greater than that which can easily be fired by hand; usually 8 ft. is taken as the maximum length. The width depends entirely on the power that the boiler is to furnish, but the sq. ft. of grate surface is commonly made such that coal can be burned at about 18 to 20 lbs. per sq. ft. per hr., and gives the desired amount of steam.

Grate Air Space: How is the percentage of air space in a grate determined?

To find the percentage of air space of a furnace grate, obtain the number of sq. ft. of grate surface. Having this, it is necessary to find the area of the openings in the grate, which may be done by finding the number of sq. in. in each opening and multiplying by the number of openings, dividing the result by 144. This gives you the area of openings in sq. ft. Dividing this last product by the total

grate area and multiplying by 100 will give you the percentage of air space.

Ratio Grate to Heating Surface: What is the average ratio of grate surface to heating surface for anthracite? For bituminous coal?

In hand-fired plants using anthracite under forced draft, 40 to 50 sq. ft. of heating surface to each sq. ft. of grate has been found to give excellent results. Where bituminous coal is burned with natural draft, there is variation in the ratio of heating surface to grate surface ranging from 50 to 78, with an average of 60. For stokers used with natural draft, conditions are much as with hand-fired furnaces, and excellent results are obtained with ratios of from 50 to 60 to 1, or projected area.

OIL BURNING

Carbon Prevention: How can the accumulation of carbon in oil-burning furnaces be prevented?

This accumulation may be due either to insufficient atomization of oil by the burners or to the air supply being insufficient or inefficiently distributed.

Careful overhauling of the burner to make sure that the tip is clean and not burned, that passages are clear and that, if a mechanical burner, it is revolving at full speed. Also, oil should be heated to at least 150°F., but not over 200° at a pressure of at least 25 lb. per sq. in.

Insufficient air will be shown by black smoke at the stack, and the remedy is to admit more air. Even with excess air, carbon may form if the air is not supplied at the right point. Some air should be admitted around the burner, the rest through the checkerwork in the furnace floor. Air spaces clear across the furnace will produce cool corners at the front end. Openings starting in front of the burner and extending to the end of the furnace will result in carbon at the rear wall.

STACKS

Stack Temperature: What are the causes of high stack temperature?

Among the causes of high stack temperature are poor combustion, insufficient air above grates, high percentage of volatile in coal, sooty or scaled tubes, poor baffling, broken baffling, improper furnace proportions, inefficient heating surface and poor water circulation.

Stack Corrosion: For a plant operating at 140 to 150 lb. pressure with fires banked at night so that the stack cools down, the 10-gauge steel stack rusted out in 4 years, rust starting apparently from the inside. A new stack is starting to rust. During the first 24 hours of operation a gummy material collected on the inside, and ran through seams to the outside where it dried out. What is it? What is its effect on steel? How can it be prevented?

Deposit such as that described is not uncommon, especially in the stacks of steamboats, and is due to the condensation of part of the constituents of flue gases due to the cooling of the stack walls by the outside air. This condensation flows down the inside of the stack, and seeps through if the joints are not tight. Its content depends on the coal burned, but is thought to be due to sulphur in the coal, the sulphur gases condensing as a yellowish deposit. This may combine with moisture to form sulphuric acid that will corrode steel.

It is possible that the deposit is some other substance due to impurities in the coal, but chemical analysis would be required to determine this point.

By making joints with each upper course inside the lower, the deposit will drip off the edges of the upper courses instead of collecting at the joints, and will either fall to the bottom or be picked up by the high temperature gases.

Steam in Stack: What will be the effect of venting steam at 120 lb. pressure into a brick stack?

If a considerable amount of steam is removed, venting it to the stack would be poor practice. It may have an eroding effect on the stack, and weaken the structure at the joints. Some cities have ordinances prohibiting this practice, and it would be better in any case to pass the steam through a reducing valve before discharging to the stack. If the amount of steam is small and is discharged as a jet upwards along the center line of the stack, and the stack is of large diameter, probably no serious trouble would be experienced. But any considerable amount of steam at such a pressure should be used for heating feedwater or some other purpose which would save the heat that it carries.

Soot in Stacks: Trouble is experienced in a plant of two 175-hp. boilers with two stacks 100 ft. by 3 ft. from accumulation of soot in the stacks. This catches fire and is a hazard to surrounding buildings. Can this soot come from the use of soot blowers? How can it be prevented?

The soot is probably from fuel containing a high percentage of volatile such as shavings or the dust and slack of soft coal. Cleaning out the base of the stack and burning the refuse in the boiler furnace will help. Another remedy would be to run a 1 1/2 in. pipe to the top of the stack, and provide a perforated pipe to give a ring spray to wash down the soot when the stacks are cooled down.

Provision should be made for draining the pipes in freezing weather. The method will not add to the life of the stacks, but it will remove the fire hazard. The stacks should be washed down regularly, as often as necessary to prevent soot accumulation.

Stack vs. Blower: A return tubular boiler has grate 5 1/2 by 6 ft. with small air spaces and burns 5 tons of coal a day. Stack is 70 ft. high, but the top 40 ft. are in bad shape. Is it better to rebuild the stack to 70 ft. or use a blower, and if so, of what capacity?

Opinions on this point vary. One engineer has had good success with under-grate blowers, but feels that 30 ft. will hardly be height enough to carry away the burned gases from above the fire. Another says that an under-grate blower with only 30-ft. stack would develop a furnace pressure which would result in rapid deterioration of the walls. Also, heat of flue gases will be sufficient to furnish the draft needed if a 70-ft. stack is rebuilt, and the operating expenses of a fan will be saved. A third man advises an induced-draft fan at the base of the 30-ft. stack because of its drawing gases away from the furnace and the flexibility of operation. There would, however, be a larger quantity of gas to handle at high temperature.

Height of 70 ft. would give 50% more fuel-burning capacity than 30 ft. But the under-grate burner would add 75 to 100% capacity.

Part 3: Boiler Plant Operation

FIRING

Clinker Causes: Name the most common causes of clinker.

The higher the percentage of iron pyrites, the more likelihood there is of clinkering. Iron pyrites is the form in which sulphur is usually found in coal. This fuses with the ash present in the coal and forms clinker. Since the fusing point of irons range from 1922°F. to 2786°F., and that of Silicon, one of the principal constituents of ash, is 2588°F., high temperatures in the fuel bed will increase clinkering. For this reason, water-cooled walls and steam injection are often used.

Improper slicing or stirring of the fuel bed increases the tendency to clinker, for if the bed of ash is raised up and brought into the hottest part of the furnace, it is almost sure to fuse into clinker.

Dead Plate Precaution: Why should scattered coal not be allowed to burn on the dead plate of a boiler?

This is because, since no air passes up through this coal (as is the case when burning on the grate), it is liable to overheat and perhaps warp and even crack the dead plate. A small amount does no harm because it simply distills off the volatile gases, changes to coke, and is pushed into the fire where the solid coke is burned when the furnace is stoked again.

Wetting Coal: Is it advisable to wet soft coal before head firing?

Properly done, wetting coal has advantages but if too much water is used it will result in loss rather than saving. The wetting needed depends on the kind of coal. Water should be added uniformly and never to the point where the coal has the appearance of being wet or being caused to cake.

Pulverized Coal Size: What is the fineness of pulverized coal?

The older practice was 95% to pass through a 100-mesh-per-in. screen and 85% through 200-mesh screen. This is now thought to be unnecessarily fine, and to involve waste of power in pulverizing. At present, 99% through a 40-mesh screen, 90% through 100-mesh, and 70% through 200-mesh is considered satisfactory.

Banking Fires: Does any bad effect on boilers or steam mains result from banking the boilers and letting the steam pressure drop every night?

To bank boilers and let steam pressure down overnight has no effect on boilers or steam mains except to cause expansion and contraction, which are usually allowed for in a properly designed boiler setting and in properly designed steam lines. Some steam undoubtedly leaks into the lines from the boiler during the night and condenses there, making it necessary to provide proper drainage for the lines and to exercise care in heating them in the morning.

Steam Jets: What is the effect of steam jets injected into the combustion space of a boiler?

They principally cause turbulence and mixture of air and combustible gases. The steam is heated up to the temperature of the gases and carries some heat up the stack as a loss. Also the heat used in making the steam is lost, and this has been known to be as high as 10% of boiler capacity. A better way to create turbulence is proper admission of supplementary air above the fire; preferably near the front of the fuel bed.

Firing Fine Wood Refuse: What rules should be observed in firing fine wood refuse (shavings and sawdust)?

1. Keep pit doors open at all times on boilers in service.
2. Control the steam pressure and main air supply with the boiler damper. Supply the necessary fuel to generate the quantity of steam required.
3. Admit air at all times over the fire by keeping the fire doors partly open. Regulate the opening of the doors to prevent the discharge of dark smoke from the stack. The stack should be clear or a light haze should leave it. Planer shavings and dry fuel require wider firedoor openings than wet, fine fuel.
4. Fire often and in small quantities rather than on large quantities and less frequently.
5. Keep the fire thin in front and heavy in rear; top of rear fuel cones should be closer to arch than top of front fuel cones.
6. Fire in front should reach about four to six inches over the dead plate at the center of firing doors. It is better to have the grates slightly uncovered at the front than to have too much fuel charged.
7. The fire at front should always appear bright and active. The firedoor opening should show bright when viewed from the firing floor; a dark door shows the fuel bed is too heavy.
8. The fire should be heavy enough in the rear to prevent vicious flame coming through at the bridgewall.
9. With coarse, dry fuel the fire should slope at rear to slightly below the level of the bridgewall. With fine, wet fuel the fire at the rear should slope to about halfway up the bridgewall.
10. Fire in the rear should be

carried heavier for coarse and dry fuels, thinner for fine, wet fuels.

11. At no time entirely close the ashpit or firing doors on a boiler in service; use the boiler damper for main control of air supply.

ASH DISPOSAL

Ash Disposal: What methods of ash disposal have proved successful for powdered fuel plants?

In most cases, the drag conveyor, pivoted bucket elevator, or gravity dumping into cars below the ash pit have proved satisfactory. The screw conveyor has been little used because of the abrasive action of the ashes, although this action must be considered with any type of mechanical conveyor.

Sluicing of ashes is rapidly gaining favor, especially in the larger plants. Ashes drop into an open sluiceway under the ash pits, and are carried by a current of water to an ash storage tank whence they are removed by special manganese steel ash pumps.

Air current conveyors are also used satisfactorily for ash handling, delivery to cars, tank or a landfill.

Wetting Ashes: Is it common practice to wet ashes before handling with other than pneumatic conveyors, and can wet ashes be stored in tanks?

When they are to be dumped directly into cars, ashes are often wet in the hoppers. For other methods of mechanical conveying it is not good practice since abrasive action is increased, and greater weight must be handled. Moistening to keep down surface dust may not be objectionable.

Ashes should never be stored wet

in a bin because they will pack, and in cold weather will freeze, making them difficult to remove.

FEED SUPPLY

Injector Action: Will an injector put hot water into a boiler?

Feed pump or injector water pressure has to be at least 6% over the highest set safety valve. Also, it must be able to deliver sufficient water at maximum forced fire of all boilers combined. This could require more than one pump formula developed boiler hp. times 0.069 equals feed pump gpm.

Feed Pump Arrangement: In a plant carrying 120 lbs. pressure a new boiler for 150 lbs. pressure is to be installed. The two steam mains will be connected by a reducing valve, but is it necessary to provide a separate high-pressure feed pump? Is there a tendency for the 120 psig. boilers to rob the 150 psig boiler of its feedwater?

If all steam is to be used by 120 psig., it would be better to run all boilers at that pressure. But presumably some steam is needed at 150 psig., the rest of the output of the new boiler to be taken by the 120 psig system, and it would be simpler to run the 150 psig. installation separately, both as to steam and feedwater. Any irregularity in reducing valve action may throw extra load on the safety valves in the 120 psig. boilers. In some states, such connections are prohibited, unless a safety valve of sufficient capacity is installed on the reducing station and set to 120 psig. Additionally a non-return valve would be required on the lower pressure boilers for protection.

Feeding the 120-lb. boilers through the regulator valves will probably work out all right if there is sufficient capacity of feed pump for all boilers. The pump pressure would have to be raised to over 150 psig but seats and discs of regulator valves may be cut rather rapidly. Hand regulation is hardly possible as conditions are too unstable. Feeding from a common main to two different pressures has been done successfully, and feedwater level regulators have been used, but operation will be more certain if a separate feed pump or an injector is used for the 150 psig boiler.

Feed Water Operation: Assuming in all cases that the water glass is correctly located with reference to lowest permissible water level for the type of boiler, and that, in the course of regular operation, the water has dropped out of sight in the glass (due to closed feed valve, stopping of pump or similar condition), is it safe practice to turn on the feed or to start to feed water into the boiler without checking the fire, provided that the water can be made to bob up in the glass by manipulating the drain valve on the glass?

If water came from the try-cock when opened, it would indicate that the water level was not below the crown sheet or danger zone. In In this case checking the fire before feeding water is unnecessary, but if no water drips from the try-cock when it is opened, no water should be fed into the boiler until the water level has been sufficiently checked. If there is any doubt regarding the height of the water in the boiler, it is best to put out the fire and investigate if water has dropped to a point where the boiler overheated. Have it inspected before refiring it.

Feed Line Water Hammer: Two boilers out of three are run at a time, fed through 1 1/2 in. branches from a 2 1/2 in. feed line. The pump is 6 by 4 by 6 in., 3 in. suction and 2 1/2 in. discharge. Running slow there is no trouble, but on speeding up water hammer develops in the feed line. What can be done to stop this?

Water hammer is often due to sudden contact of cold and hot water, or may be from steam binding on the pump. The pump on speeding up may have a jerky motion due to increased slip.

If all water is the same temperature, the first cause is eliminated. If not, slower mixing in a suction or discharge receiver is the remedy, or perhaps dropping the temperature of the feedwater.

If feedwater comes to the suction hot, it may produce steam when pumping fast. Bringing the water to the suction under head is the remedy.

If slip is the cause, overhauling the pump to prevent leakage past the water pistons or valves or in the steam end will help. Also, receiver chambers attached to suction and discharge ends, especially if piping is long, will steady the flow. A larger pump may be needed, although the size mentioned should be sufficient for 300 boiler horsepower (which is about 21 cpm feedwater).

BLOWDOWN

Blowdown Period: A boiler is operated 10 hrs. a day with little makeup and practically no scale or dirt. When is the proper time to blow down--during the day, when you have a a full head of steam, or before firing up in the morning, when you still have from 20 to 40 lbs. of steam left from the previous day?

The best time to blow down a boiler is after the boiler has been standing idle for some time so that the water has had time to settle. Therefore, blow down before firing up in the morning.

Measuring Blowdown: Where is it desirable to meter the blowoff from boilers and how can it be accomplished? Weighing is impossible because of evaporation, and maintenance of orifices or throats would be excessive because of erosion by the scale. Could gauge glasses be calibrated to show by fall of water level the weight blown down?

The water level would require careful calibration and might be inaccurate, for water level in the boiler and glass are often not the same, the difference depending on the construction of the boiler and the rate of firing.

One method would be to install a blowoff tank of such capacity that at least one boiler could be blown down (without overflow) with a cooling coil or heater in the vent to condense vapor. Contents of the tank could be measured before and after blowoff with a calibrated stick inserted through a gate valve on top of the tank. This would be somewhat expensive to install.

Weighing is more accurate than might be thought. A good approximation is obtained by multiplying the weight obtained by 1.15 for 100 lbs. pressure, 1.19 for 150 lbs., 1.22 for 200 lbs., 1.25 for 250 lbs.

Emptying Boilers: In one plant boilers are emptied by running pressure up to 90 lbs. to blow the tubes, then blowing all the water out at 30 lbs. pressure. Is there an advantage in this method?

It is undesirable to empty a boiler under pressure. After blowing tubes the boiler should be allowed to cool before draining

until the pressure is practically zero. If emptied under pressure, plates and brickwork are hot, and this tends to bake scale and sediment on the plates so that they are difficult to remove. On the whole, no time is saved, for if cooled before emptying, most of the sediment can be washed out. Draining under pressure endangers plates and setting by strains from too rapid cooling. Remember that when a boiler is shut off (no fuel) the boiler is still steaming because of the hot brickwork. Draining the water below a safe level will result in burning the drain and tubes.

CLEANING

Scale Removal: What is the best method of removing thick scale from the tubes inside a Manning boiler? Would you recommend the use of kerosene?

If the scale is thick it is best to remove the tubes, clean them thoroughly, and return them.

The scale portion can be reached when dry, and kerosene, among other chemicals, will help when it is sprayed over the surface. Kerosene has no chemical action on the scale; it merely penetrates between the scale and the plate and loosens the scale, but it may remain in place until the boiler is put under pressure. Therefore, when kerosene is used the boiler should be shut down within four or five days and thoroughly washed out. If kerosene is used care must be taken that there is no open flame in the vicinity. A boiler chemical company is likely to recommend a better method than the use of kerosene.

It is also advisable to have a chemical analysis of the water made to determine what causes formation of the scale.

necessitates taking the boiler out of service for repairs when the leakage from the crack is sufficient to interfere with the fire or to cause corrosion of the shell. The boiler should be taken out of service immediately if the crack is found to extend through a rivet hole and into the plate in a direction opposite from the caulking edge. The cause of this condition can be age or a misuse of a boiler, and can result in a rupture, which is known as boiler explosion.

Fire Crack Repairs: How can a fire crack in front of a rivet on the girth seam be repaired?

A "V" groove may be chipped along the crack, the rivet hole countersunk, and a new rivet driven to fill both countersink and groove. The groove may also be chipped and then filled in by electric or oxyacetylene welding, and a new rivet driven in place of the old one. Repair is not essential if the crack extends from rivet to caulking edge and there is no leak. If the crack extends away from the caulking edge, but not beyond the edge of the inner sheet, the weld method is alright. If the crack is beyond the edge of the inner sheet a patch must be riveted on and caulked.

Cement for Pitting: Will cement wash help to prevent pitting of boiler drums?

It has been used to advantage. The wash was made of clean Portland cement mixed in water to a thin consistency and applied with a paint brush after thorough scraping and wire brushing. All small blisters were removed. Two coats of the wash were applied to two 48-in. by 18-ft. drums at first, and worked so well that others in the plant were afterwards treated. The boilers carry 100 lb. pressure and have shown no pitting in three years since

the treatment. The same wash is used for elevator and hot-water storage tanks.

High-Pressure Gaskets: What gasket material and application method is best when connecting pipe lines to high-pressure boilers?

Most engineers prefer a very thin gasket, not over 1/16 in. thick. Corrugated metal with or without corrugations filled with asbestos has been found satisfactory.

Care should be taken that the face flanges are smooth, parallel, and in line. The least pressure on gaskets that will prevent leakage is best. Much present day practice favors ground joint flange faces with no gasket and edges of the flanges welded, especially where very high pressures are to be used.

Leaky Handhold Plates: In water tube boilers trouble has developed from leakage around the handhold plates. Various gaskets have been tried. Lead works best, but even then there is considerable leakage at times, especially when firing up or taking boilers off the line. Is there a remedy?

If in use for some time, perhaps the seats have been corroded. First, be sure that the seats are thoroughly cleaned of old gasket and scale. If corroded, they should be faced off to a smooth, true surface. Using lead gaskets the bolts should be taken up the first time or two to reduce the pressure, since there may be slight loosening of the bolts due to thinning of the gaskets when steam pressure is raised. This should eliminate leakage, but the best kind of gasket depends somewhat on the water used and can be determined only by experiment. An old gasket should be thoroughly cleaned, and scratching the surface on which it sits should be avoided. A light graphite lubri-

Oil in Boilers: What is the best method for removing oil from boilers?

When possible, after emptying the boiler pour several pails of soda ash solution into it. Fill the boiler with water, and fire in low fire position for several hours, not allowing the steam pressure to exceed 15 lbs. From time to time let out part of the water and replace with fresh water. Continue for about 12 hours. Then cool boiler, empty it, and rinse with fresh water. Blow down the boiler at intervals using a surface blowdown if your boiler is equipped with one.

Cooling Boiler: How would you cool a horizontal return-tubular boiler in the shortest possible time?

To prepare a boiler for cleaning, the fire should be allowed to burn as low as possible and then pulled out of the furnace, the doors being left slightly open and the damper wide so that the walls may gradually cool. Because of the injurious effects on the plates and the seams of the shells and the brickwork of the setting, it is just as bad to cool a boiler off suddenly as to fire it up too quickly. After the boiler has become sufficiently cool and no more pressure is indicated by the gauge, the blowoff valve may be opened and the water allowed to run out. If operating conditions allow it, a boiler should be left for 18 to 36 hrs. or to about 100°F. after being cut off from the line, after which it may be emptied, opened and entered.

Boiler Skimmer: How does a boiler skimmer operate?

When water enters a boiler it is likely to carry foreign matter that rises with circulation and forms a scum on the surface. For a time this scum will float, but after a while it will pick up mineral matter, sink and settle on tubes and shell, forming a scale. To remove the scum continuously before it sinks, and to transfer it to a reservoir whence it can be blown off with little waste of water is the function of the skimmer. It maintains a down-flow circulation from the surface of water in the boiler through the outside reservoir and back to the bottom of the boiler. The scum thus is removed in the reservoir, which should be blown frequently to remove the deposit, say every 3 to 5 hrs. The skimmer will not take the place of water softening and blowing down the boiler, especially if feedwater contains much mineral matter, but it will reduce the rate of scale formation and may even prevent new scale forming.

REPAIRS

Blistered Boiler Plate: How would you repair a blistered boiler plate?

Blistered boiler plate is ordinarily found in the fire sheet of the horizontal tubular boiler. The outer skin of a blister may be chipped away. If of no greater thickness than 6% of the plate thickness for quadruple riveted butt joint, and no greater than 14% of the plate thickness for triple riveted butt joint, no further attention is necessary unless the blistering continues. If the lamination that causes the blister travels at at angle through the sheet, as can be determined by the variation in the wall thickness of the blister, patching may be necessary.

Fire Cracks: Is a fire crack dangerous enough to discontinue the boiler from service?

Fire cracking in the girth seam of a horizontal tubular boiler cant or high temperature grease should be applied between gasket and metal.

Tube-Sheet Bulge: If you find the segments between tubes in the rear tube-sheet to be bulging outwards slightly, what is the probable cause, and what should be done to remedy it?

If the segments between the tubes in the rear tube-sheet bulge outwards slightly, it is likely that the tube-sheet was not flat originally. If the tubes are sound in a bulged tube-sheet, it is best not to attempt to flatten the tube-sheet as harm will result. If the tubes should be renewed, the tube-sheet may be flattened after removal of the tubes by proper hammering and suitable backing to strike against. Shut the boiler down at once for inspection. A rupture will result in a boiler explosion.

Tube Renewal: What precautions are needed in putting a new tube in the bottom row on a B&W boiler? How would you get the old one out? What kind of a cutter is used?

The method ordinarily used in cutting out a tube is as follows: First, two slots about 3/4 in. apart are cut in the bottom of the tube with a flat ripping chisel to the inside face of the seat. This narrow strip is then bent with what is known as an oyster knife. A round nose chisel is then used for cutting the tube inside the seat for a distance of 4 to 6 ins. The slotted portion of the tube is then turned in, and the diameter of that portion originally expanded to make a seat is decreased so that the tube may be drawn through the tube hole. When cutting out a tube great care should be taken not to injure the seat, that is, not to cut a groove or slot across the seat. All cuts should be made clear of the seat.

Short Tubes: If tubes ordered for a return tubular boiler came 1/4 in. short, what could be done with them?

If boiler tubes were ordered with the intention of beading them and they came short by 1/4 in., they could still be used by rolling them and flaring the end out with the peen of a hammer because many boilers operate regularly with flared tubes.

In water tube boilers the tubes are seldom if ever beaded. In railroad practice flaring is becoming more common. Tests have been made with boilers in which the tubes have been set by the several methods, and it was found that a flared tube withstood a slightly higher pressure than a beaded tube when tested under hydraulic pressure to destruction.

Tubes Loose in Heads: A 125-lb. combination return tubular and water tube boiler was retubed a year ago because tube leaks started in the rear head, and the old tubes (11 yrs. old) were too brittle to roll. The new tubes have developed leaks three times, and are starting to leak again. Tubes, heads and breeching are kept clean and free of scale. Cooling strains are avoided, but furnace temperature is excessive due to overload on the boiler. The rear wall was moved back 32 ins. from the rear head, and a flat suspended arch put in to replace the curved arch to avoid scrubbing of flames on the rear head. Leaks still appear. What is the remedy?

Strains at the head may be caused by cold water feed into the boiler near the back end. That the old tubes were brittle after 11 years indicates that excessive heat or something in the water caused the tube metal to crystallize. Care should be taken in rolling because too much rolling on one tube may set up leaks in others next to it.

Frequent rolling thins the tube and stretches the head so that joints cannot be kept tight.

Leaks are generally due to over-heating or local cooling, and these may be caused by even slight local accumulations of scale at the joints that may not be noticed on casual inspection.

Superheater Repairs: What mishaps occur most frequently to superheaters, and how are repairs made?

Burning out from too small a steam flow to carry away the heat, and "scaling" due to deposit from wet steam are the chief troubles. The tubes are of small diameter, and usually curved so that they cannot be cleaned out. Removing tubes and installing new ones is the only method of repair.

Replacing Glass Gauge: What precautions should be taken when putting in new gauge glasses on a cold boiler?

When putting in a new gauge glass, care should be taken to see that no unnecessary pressure is exerted by the stuffing box glands. It is also good to have the glass a trifle shorter than the distance between the bottoms of the stuffing boxes, so that as the glass expands due to heat of the steam and water, it will not be placed under a strain between the bottoms of these boxes. The glass should not touch metal at any point. You should be certain that all broken glass and washers are discarded, and that the boiler is not plugged; then install the new glass.

OPERATION DETAILS

Boiler Efficiency Approximation: Without charts is there any way of telling at what efficiency a boiler is running?

The approximate efficiency of a boiler may be obtained by measuring the amount of water and coal added between hourly periods and obtaining averages. The amount of fuel may roughly be obtained by counting shovelfuls, the amount of water by using a counter on the feed pump. These give only an approximation, but may be used for comparative purposes.

Cut-in Pressure: At what pressure should a boiler be cut in with other boilers?

The pressure on the boiler that is to be cut in should be the same or a little above that of the header. It should not be more than 4 or 5 lbs. above the header pressure. The steam valve should be opened slowly.

Heat Loss: How many Btu and equivalent in coal is lost on the bare surface per sq. ft. of a boiler carrying 110 lbs. pressure with an atmospheric temperature of 75°?

Radiation losses from bare surface areas vary with the kind of surface and the temperature difference. From the drum of an uninsulated boiler carrying a 110-lb. gauge pressure and surrounded by atmosphere at 75°, the radiation is about 3.17 Btu per sq. ft. per deg. temp. difference per hr., or taking the temperature difference at 270°F., the loss would be 85 Btu per sq. ft. per hr. With coal of 10,000 Btu and boiler efficiency of 75%, the loss per sq. ft. per hr. is approximately 0.15 lb.

Stay Bolt Broken: How can a broken stay bolt be located?

If the stay bolt is hollow it will leak if broken. If it is solid the best way to find out is to go inside and, tapping lightly with a hammer, you can tell by the sound whether it is broken or not.

Priming, Foaming and Corrosion: What causes priming, foaming and corrosion in a boiler, and what can be done to avoid these troubles?

Priming is the pulling over of water with the steam due to the intense overload or excessively rapid steaming. It is best remedied by reducing the load.

Foaming is due to impurities in the water. These impurities boil to the surface, and if a surface blowoff is available it will help relieve the situation. It is also good to open up the bottom blowoff, and feed in a fresh supply of water to reduce the density.

Corrosion is due to the acidity of the water or to oxygen in the water. If the water is acid it is best to get another source of supply. If oxygen is causing the trouble, deaerate the supply or use some mixture that will absorb the free oxygen. Hydrated alkalinity to 200 ppm will take care of the acidity in the feedwater.

Water Line Corrosion: What coating can be used to reduce or prevent corrosion in a boiler on a strip from 8 in. above normal water level to 14 in. below? Spring water free of impurities is used.

One treatment that gave success was a coat of Portland cement and boiled oil, mixed thick enough to be applied with a stiff brush, followed by a coat of one part graphite to three parts cement mixed with boiled oil.

Another mixture used was boiled oil and zinc oxide. The surface was sand-blasted to clean off all rust, the coating applied with an air brush, allowed to dry two days, and a second coat applied.

Application of zinc coating by the Schoop metal spraying process on a sand blasted surface has also been successfully used.

Oil in Boiler: On recent inspection, tubes, and the front head and through stayrods above the tubes, were found coated with 1/64 in. of oil. Stayrods below the tubes were loose, bottom tubes bent up against each other, and some beads cracked at the rear head. At the bottom the shell was bowed up, showing 1 1/8 in. bending of the center course. Some tubes pushed out through the heads, but most of them held and buckled. The bottom of the boiler was clean inside. The boiler is 84 in. by 20 ft., 9/16 in. plates in three courses, 118 4-in. tubes, 3 through stays below and 6 above the tubes. Support at four points. Pressure was 125 lbs. and the fuel bagasse. What was the cause of the distortion?

Analyzing the conditions, most of the oil probably gathered on the bottom of the tubes, which became hotter and weaker than the tops so that the tubes bend upwards. Since the shell was clean it did not get as hot or expand as much as the tubes, hence the tubes shoved through the heads or buckled. This would tend to pull in the lower part of the heads. Also when cooling down, the tubes would contract more than the shell, increasing this tendency and the bottom of the shell would buckle in the direction of least resistance, that is, upwards. Upward bending of the shell would be induced by the lower parts of the heads drawing together and tilting the end courses with them, also by the center course being, as is usual, smaller in diameter than the end courses.

Oil Removal: When boiling soda ash in a boiler to remove any oil that may be present, would you leave the manhole plate out on top or not, and why?

In boiling soda ash in boilers it is usually allowed to act with a pressure not exceeding 15 lbs., and thus it is necessary to keep manhole closed.

Local Heat: What effect does constant, intense heat have on one point of a boiler? Is there danger of buckling or rolling due to the concentration of heat?

No ill effect such as buckling or rolling is experienced when intense local heat occurs in a boiler if freely circulating water is on the cooling side, provided there is no dirt in the tubes, including oil or grease.

Increased evaporation at a highly heated local spot occurs. This may cause a greater accumulation of scale at that point that may cause bagging, but no trouble occurs if the plate is kept clean and water circulates freely on the cooling side.

Electrolysis in Boilers: What current and voltage produces electrolytic corrosion in boilers?

Electro-chemical corrosion may be produced by such substances as acids, greases, or oxygen in the feedwater. Galvanic action may be set up by the differences in composition of the metals in boiler structure, setting up minute battery action. The voltage depends on the materials involved, but is small, in the order of 0.1 to 0.2 volt.

Corrosion is cumulative and may result from a small current flowing for a long time, or a large current flowing for a short time. The amount of corrosion depends on the ampere hours, and voltage enters only as a means of setting up the current.

Fitting boilers with zinc plates has been a practice used to overcome such corrosion, and the currents localize from zinc to steel, thus corroding the zinc instead of the boiler. The zinc must be kept clean and renewed when corroded because a coating of scale or oxide reduces the effective action.

Another method is to supply an outside electromotive force of about 6 volts from a battery or small generator between plates suspended inside the boiler structure, so that these plates will be eaten away rather than the boiler. Current of 9 to 10 amperes is found sufficient.

Embrittlement: What is caustic embrittlement of boilers, and how is it caused and prevented?

Embrittlement refers to an apparent crystallization and subsequent cracking of the plates at joints and around rivets due to an excess of caustic solutions in the feed-water. Cracks usually occur at the joints between rivets, and in the rivets at points reached by the water, without relation to the stresses due to pressure in the boiler. Brittleness is shown under shock or fatigue of the metal in areas localized at the points, even though little indication of brittleness may show under static stress. Brittleness partially disappears in time when the boiler is taken out of service, especially if the joints are heated at temperatures lower than those required for annealing. A surface indication is plentiful deposit of a black powder, which tests chiefly as iron oxide.

The cause is a substance that does not attack the boiler when in highly concentrated solution, but does in the higher concentration of a deposit that builds up at the joints. This eliminates acids and such substances as calcium and magnesium carbonates or calcium sulphate. High conductivity will increase the corrosive effect.

Substances which are probable causes of embrittlement are sodium hydroxide and possibly chlorides of calcium and magnesium. Cracking from embrittlement has chiefly occurred in 1) caustic soda works where the water was nearly pure except for sodium hydroxide; in 2) boilers using artesian well-water; and in 3) marine boilers using a compound made largely of sodium carbonate and caustic soda, which might turn to sodium hydroxide.

Sodium carbonate can, however, be used as water treatment unless this results in a high concentration of sodium hydroxide, which generally will not happen. Cracking from embrittlement may lead to explosions, but plenty of indication is given before the danger point is reached. Leakage appears and caulking does not keep the joints tight. Inspection will often show the accumulation of a deposit, but rivets should be taken out, holes cleaned, and in some instances, plates separated and examined. Cracks develop first in the center of the joint and on surfaces where plates are in contact so that, unless a thorough examination is made, the trouble may not be detected.

One remedy is the neutralization of the sodium hydroxide by acid, but this has dangers from careless use of too much acid. Another remedy is the use of magnesium sulphate, which will form some scale and should be done with care.

Keeping down concentration of sodium hydroxide by frequent blowing down, avoidance of localized stresses and of forcing will reduce the tendency to embrittlement.

Safety Valve Setting: How can a safety valve be properly reset when the blowoff pressure is to be raised from 105 to 120 lbs.? Should you reset the valve yourself, or have a boiler inspector do it?

In some states it is unlawful for an operator to change the setting of a safety valve. It is best to take the matter up with the local boiler inspector, or with the Board of Boiler Inspection at your state capital.

For pop safety valves changes of adjustment must be made by trial. Spring loaded safety valves differ in design detail, but all are adjusted by changing the tension on the spring. If the valve is to blow at a higher pressure, the screw that regulates the amount of compression on the spring must be screwed down (the actual amount being determined by trial). The cap over the top of the adjusting screw is first removed, and the screw turned down a small amount, the number of turns being noted. Only a fraction of a turn will probably be needed. The boiler pressure is then raised, and the pressure at which the valve blows noted. This will give an indication of the approximate amount the screw must be turned for the pressure desired. The exact amount must be found by trial. All safety valves on boilers are set at or below the maximum allowable pressure, and also at forced full fire prevent.

Safety Valve Blocking: How should the safety valve be blocked or held closed when making a hydrostatic test on a boiler?

The safety valve is closed by means of a gauge. This gauge is a yoke that clamps to the body of the valve, and carries a set screw adjusted so that the spindle of the valve is prevented from rising.

Safety Valve Adjustment: The pop safety valve does not stay open until the pressure is reduced, but keeps popping at short intervals.

What can be done to remedy this?

If the pop safety valve "flutters," adjust the blowdown or the blow-back ring from 2 to 8 lbs. before closing and never more than 4% of set pressure. If this ring is screwed down too much, the valve is too sensitive.

Non-Return Valve Chatters: Four 150-hp. return tubular boilers at 140 lbs. pressure are set in a row and connected by 6-in. branches to a 10-in. header, 40 ft. long. In each branch is a gate valve and an automatic non-return valve. Steam is supplied to one 450-hp. and two 175-hp. uniflow engines through 6-in. branches. Ordinarily, two boilers are used at one time. With Nos. 1 and 2 steaming, non-return on No. 2 chatters. With Nos. 3 and 4 steaming, No. 4 valve chatters, no matter which engines are running, but worse at light loads. How can this be overcome?

Chattering is due to steam impulses from the reciprocating engines, and the noisy valve probably is the one nearest the engine in each case. The dash-pot piston must be adjusted to a good fit, the vent port not too large, and the valve disc should telescope the seat at least one inch. Make sure that the piston ring has an easy fit in its groove. Drill a 5/32 in. hole through the piston.

Part 4: Fuels and Combustion

HEAT VALUES

Heat Values of Fuels: What are the comparative steam-making values of coke, anthracite, and bituminous coal, hand-fired for moderate pressures?

Calorific values of coke range from 10,700 to 11,900 Btu per lb.; anthracite from 11,700 to 13,300 Btu. Because of efficiency of burning and effectiveness of firing, the average steam per pound will probably not be greatly different for the three fuels, but for high steaming rates bituminous shows greater evaporation per pound than coke or anthracite.

Wood Fuel Value: What is the equivalent in coal for the various kinds of wood used as fuel?

Approximately 1 cord of ash, beech, birch, cherry, elm, hickory, locust, hard maple, or long leaf pine will equal 1 ton of coal. For light woods such as douglas fir, red gum, western hemlock, soft maple, short leaf pine and sycamore, 1 1/2 cords will equal 1 ton of coal.

Sized Coal vs. Run-of-Mine: Is it good economy to use sized coal for stokers rather than run-of-mine?

Tests seem to show a considerable gain in both power and efficiency from using sized coal.

Oil Heat Content: Why is the heat content value of oil so much higher than that of coal?

Oil has the higher percentages of carbon and hydrogen. For a 13,000 Btu coal, carbon is 60 to 70% and hydrogen 5%. For oil, carbon is 85% and hydrogen 12 to 13%, which would give a heat content of 18,500 to 19,500 Btu. Pure carbon has a heat value of 14,000 and pure hydrogen 53,000 Btu per pounds. Multiplying the percentage of each in a fuel by its pure heat value, the sum is the value of the coal.

Gas Burners: How should natural gas burners be installed under a 66-in. by 16-ft. boiler, now arranged for hand firing with coal? How large should the combustion space be, and what kind of baffling should be used?

Natural gas is best burned by the use of a large number of small burners, each being capable of handling approximately 30 rated horsepower. This obviates the danger of any laning or blow-pipe action of the gases that might be present if large burners were used.

Where natural gas is the only fuel the burners should be evenly distributed over the lower portion of the boiler front. If this fuel is used as an auxiliary to coal, however, the burners may be placed through the fire doors.

A large combustion space gives the best results, say 2 cu. ft. per rated horsepower. A checker-work wall is sometimes placed in the furnace about 3 ft. from the burners to break up the flame, but with a good type of burner and furnace this wall is not necessary. Good results are secured with burners inclined slightly downward toward the rear of the furnace where the gas is burned alone and there are no coal-fired grates.

TESTS AND COMBUSTION

Combustion Processes: Explain the entire process of combustion as developed in making a coal fire.

In the process of the combustion of coal, the coal is first heated by burning of other lighter fuel or coal that has been ignited; distillation of the volatile gases then takes place; burning of those gases above the fuel bed follows; and finally, the coke on the grate is burned. If the fuel bed is too thick, producer action may take place. In this case the fixed carbon in the lower part of the fuel bed, where the oxygen supply is adequate, burns to carbon dioxide. As this carbon dioxide rises upward through the thick fuel bed,

it may be reduced to carbon monoxide by picking up another atom of carbon, thus forming two molecules of carbon monoxide. To burn this CO to CO_2 requires the admission of air over the fuel bed.

Combustion Chamber: What is the first essential of a combustion chamber?

The first essential of a combustion chamber is to burn the gases that have been distilled from the coal on the grate by maintaining the temperature above the point of ignition of the gases and mixing them thoroughly with air.

CO_2 Determination: How is CO_2 determined and what is its relation to combustion?
CO_2 most commonly is determined by the Orsat apparatus or other gas analyses. A representative sample of gas is drawn from the uptake by means of a water bottle, and a definite volume passed to a measuring burette. This measured amount is then passed over a caustic potash solution that absorbs the CO_2 and the volume of the remainder is measured. The volume lost is CO_2 and calculated as a percentage of the original volume measured.

For complete gas analysis, the O is then similarly determined by passing the remainder of the measured sample over a solution of pyrogallic acid and noting the volume lost. Then CO is determined by passing over curprous chloride.

Special CO_2 indicators are made that use only caustic potash, and do not determine either O or CO.

For complete theoretical combustion the percentage of CO_2 would be about 21, but such a condition is impossible, and 17 to 18 is the maximum obtainable. Good practice will run from 13 to 16%. Lower percentages should be investigated, although with some furnaces 12% is the highest possible.

Flue Temperature and CO_2: Why is flue gas temperature high with low CO_2 and vice versa?

Percentage of CO should also be considered. With low CO_2 and high CO there is incomplete combustion and insufficient air so that furnace gases do not reach their highest possible temperature. Increasing the air supply will increase CO_2, decrease CO and raise the furnace temperature up to the point that CO disappears. More air above that point means a great loss of Btu out of the stack, and a greater amount to be heated with no more heat available, hence a lower temperature of the furnace gases. Also CO_2 will test lower because the gases are diluted with too much excess air. Why then should low excess air and high CO_2 give low flue gas temperature?

Hotter gas entering the tubes results in faster transfer of the heat to the water because of greater temperature difference between gases and water in the tubes. After gases have passed over a certain amount of tube surface, this temperature head is decreased but the rapid transfer during the early stages is sufficient to result in lower temperatures of the flue gases than that obtained when furnace temperature is lower.

Best Percent CO_2: What percent CO_2 will give the best efficiency operating a 264-hp. water-tube boiler with underfeed stoker at outputs of 100 to 200hp.? At 400 hp., CO_2 is 17%, but below rating dampers have to be partly closed and, admitting air sufficient to prevent burning the stoker bars, the CO_2 goes down to 12%.

For the outputs given, 12% seems about right, especially if analysis shows not over 0.2 of 1% CO in the gases. As to the 17% at 400 hp. such a figure is reasonable, but CO should also be checked since high CO_2 readings may accompany large CO because of insufficient air supply at high outputs. If high CO_2 is accompanied by low or no CO and normal 0 readings (not over 6%), conditions are excellent. The 12% CO would indicate about 75% excess air, which is not unusual at low boiler outputs.

CO_2 for Coal and Oil: Why is the maximum possible CO_2 higher with coal than with oil firing?

Possible CO_2 depends on the composition of the fuel. For pure carbon it is 20.9%, assuming complete combustion with no excess air. All fuels contain hydrogen, most of them oxygen and some sulphur. The hydrogen burns to H_2O water, and sulphur to SO_2, but the SO_2 is absorbed by the caustic potash and shows on test as CO_2. This error is slight, because only a small amount of sulphur is involved, and 1 lb. of sulphur requires only 1 lb. of oxygen for combustion as against 8 lbs. oxygen for 1 lb. of hydrogen, and 2.67 lbs. oxygen for 1 lb. of carbon.

Maximum CO_2 is given by the formula in the previous answer, which shows that the greater the C, the greater the possible CO_2, while the greater the H, the less the possible CO_2. Fuel oils usually contain more hydrogen than coals, hence their possible CO_2 is smaller.

FLUE GASES

Soda Ash Washing: Soda ash is being used to wash flue gases of approximately the following composition: CO_2 - 7.5%; C_2 - 7.5%; CO - 0%; N_2 - 15%. The flue gases are produced by burning natural gas under boilers, and are being blended with the vapors from a gasoline plant to make a gas of about 900 Btu. The purpose of the washing is to remove objectionable sulphur and nitrogen compounds, particularly the latter.

Since sodium carbonate absorbs CO_2 the question has arisen: Will the soda ash soon cease to be alkaline from absorbing CO_2, or will there be enough air in the flue gases to "reclaim" the soda ash? Will the soda ash under the conditions form a corrosive acid? In addition, these flue gases are compressed to above 200 lbs. and cooled by water jacketing before being washed with the soda ash solution. Will not this moist gas with oxygen present be corrosive to ordinary steel or iron line pipes?

Absorption of CO_2 by sodium carbonate solutions will not proceed beyond conversion of the carbonate to an equivalent amount of bicarbonate. Sodium bicarbonate is alkaline and is capable of neutralizing the acids in the flue gases. Hence the fact that the flue gas contains CO_2 has no effect on the efficiency of the solution in removing these acids.

Absorption of nitric, sulphuric and sulphurous acids by the soda ash solution will proceed until the sodium carbonate or bicarbonate are all converted to sodium nitrate, sulphate, and sulphite. If the passage of flue gas through the solution is continued after neutralization of the soda ash is complete, the solution will become acid. Such a solution would be corrosive, but its corrosiveness is due, not to the soda ash, but to the presence of acid in the solution.

The question may refer to the flue gas before or after scrubbing. The flue gas prior to washing will be corrosive to iron and steel pipe due to the presence of nitric and sul-phuric acids, and to a lessor extent because of the presence of oxygen and CO_2 along with moisture. After scrubbing, the amount of corrosion would be materially less. We would, however, expect washed flue gas to be somewhat corrosive, due to the fact that the moist gas will contain oxygen and carbon dioxide.

Flue Gas Loss: What is the formula for the loss of heat in flue gases?

127 minus L equals 88.08 x T divided by CO_2, where L is loss in Btu per lb. of carbon burned, T is temperature difference between flue gases and atmosphere, and CO_2 the percentage of carbon dioxide in flue gases.

Draft in Battery: In a battery of four boilers which has the best draft and why?

The best draft is on the boiler nearest the chimney because the draft is reduced on the other boilers by the amount necessary to overcome the friction of the smoke pipe. Frequently it is attempted to overcome this somewhat by using a tapered smoke pipe that is larger near the chimney than at the far end, thus reducing the friction of gases in the pipe.

Part 5: Feedwater

TESTS AND TREATMENT

Test of Hardness: Using standard soap solution for testing hardness of feedwater, 1 to 100 cc. of water equals one grain hardness per gal. Solution recently bought is labeled "1 cc. equals 0.001 gm. calcium carbonate." How is correction made to measure grains hardness per gallon?

The first solution mentioned is American standard, the second French standard, which means 1 cc. solution equals 1 gram hardness per 100 grams of water. This second means that 1 part solution to 100 parts of water equals 1 part hardness in 100,000 parts water and parts per 100,000 divided by 1.714 equals grains per U.S. gal. If you use 8 cc. to soften 100 cc. of water, it shows 8 parts hardness per 100,000 parts and 8 divided by 1.714 equals 4.66 grains per gal.

Tri-Sodium Phosphate: Using tri-sodium phosphate for conditioning feedwater, an oxalate test shows a boiler free of calcium, but adding even 10 lbs. additional phosphate to the 25,000 lbs. of boiler water shows no excess on test. What becomes of the excess?

If calcium carbonate is in the water, adding the sodium phosphate gives a precipitate of calcium phosphate and, by a double reaction, caustic soda and carbonic acid. The two last mentioned dissolve in the water. If excess phosphate is used, a double reaction produces caustic soda and orthophosphoric acid, which are soluble. Strong alkaline reaction of the caustic soda may make the determination of the phosphate content difficult.

Possibly old scale in the boiler is using up the excess phosphate, and after the old scale is gone the excess phosphate will show.

An oxalate test for calcium is not recommended because it does not detect small amounts of calcium. Also, the tendency of calcium to form scale depends on the type and concentration of acid content in the water so that with certain acid radicals, scale might be formed by a calcium content too small to show on oxalate test.

Soda Ash Use: If you had a boiler badly scaled, how much soda ash would you use? That is, is there any rule in regard to so much per horsepower? Or do you have to experiment to find the right amount?

It is found that a small quantity of soda ash will act with good results on large quantities of feed. If too much soda ash is used it is liable to promote priming, with all the dangers as well as the inconveniences that accompany it.

The best method is to connect the feed pump or injector to a soda tank so that a supply of soda solution can be drawn at regular intervals. The proper amount is usually determined by experience, but ordinarily varies between 1 and 2 lbs. per day for the average boiler. The soda will not injure the boiler unless it is impure and contains acids.

Feedwater Treatment: With feedwater containing considerable scale-forming material, what can be done to the boilers themselves to keep them in good condition?

The best way to handle any boilers with bad feedwater is to treat the water before it enters the boilers, pass it through a feedwater heater and purifier, then see that the water is fed in so that it will first pass to the mud drum or to the rear of the boiler, where the sediment will be deposited and can be blown out. The point is to get the scale-forming material deposited in a soft mud instead of in a hard scale, and to arrange the circulation so that this mud will be thrown down as far as possible from the hottest part of the boiler, so that it will not bake on the plates.

Zeolite Process: What is the zeolite process for softening water?

Water is passed over a bed of zeolite gravel, which is made up of complex silicates of aluminum, sodium, potassium, calcium, magnesium and other salts. Sodium is given to the water in place of calcium and magnesium, the sodium salts being non-scale forming. After a period of use the zeolite bed should be regenerated by passing it through a strong brine of common salt that will restore the sodium to the zeolite.

Removing Cylinder Oil: Cylinder oil comes to certain boilers emulsified with condensate so that filtering through excelsior does not remove it. What can be done to overcome this?

One difficulty may be a compounded cylinder oil that will emulsify with water, and can be removed only by longer settling than is possible in the power plant cycle of operation. Also, probably too much oil is being fed to the engine cylinders, the tendency in most plants. The remedies are to use a straight mineral oil, and reduce the supply to the minimum needed for correct lubrication.

Installing an oil extractor in the exhaust line with drains and traps and filters in the feedwater line should remove such oil as then gets in the exhaust. The oil extractor or separator should be some 10 ft. from the engine cylinder.

HEATING

Heating Feedwater: Does increasing the feedwater temperature raise boiler efficiency?

There is no effect on boiler efficiency, but the boiler has less work to do in making a pound of steam. If feedwater is heated by waste heat of exhaust steam or flue gases, there is a saving in fuel and a gain in plant efficiency.

Open vs. Closed Heater: Does an open or a closed heater give greater savings?

Saving in plant efficiency is about 1 percent for each 11°F. that feed-

water is heated. Open heaters in which steam comes in direct contact with the water will raise temperature within 1 or 2° of steam temperature, if there is adequate steam supply. Closed heaters with 1/3 sq. ft. of surface per boiler horsepower will heat water to not nearer than 15° to 30° of steam temperature, depending on the amount of steam and the rate of flow. The closed heater would then be 1 to 2% less effective in raising plant efficiency. If an open heater is used the steam must be of such condition that the condensate will be suitable for boiler feed.

Heater Efficiency: Running a 24 by 54 by 36-in. cross compound engine at 40 rpm, 26 in. vacuum, 500 hp is developed for 15 lbs. steam per horsepower-hour. Boilers evaporate 230,000 lbs. water in 24 hrs. Using a low pressure heater between engine and condenser, feedwater temperature is raised from 180° to 200°, but leaks cause vacuum to drop to 23 in. Is the heater an economy?

First, overhaul the heater and stop the leaks. Indicator cards will show the loss of horsepower with vacuum drop to 23 ins., and this must be compared with the heat saved by raising feedwater temperature. In general, cost saving is about 1% for 10° rise in feed temperature, or the total saving would be 2% for the case mentioned. For the type of engine given, loss in operating economy from 3 ins. drop in vacuum is some 5%. The net seems to be a loss of 3%. But again, why not stop heater leaks and maintain the vacuum?

Heater Surface Needed: What surface will be required for heating 1500 gals. an hr. of water from 35° to 200° with steam at 5 lbs. pressure?

If steam is free of oil the best

way is to mix the steam and water,
which does away with surface required and gives 100% transfer. To
heat 1500 gals. through a range of
200-35 or 165° will require 1500
x 8.33 x 165 or 2,061,675 Btu an
hr. Heating by surface heater
the condenser will discharge at
say 212° with heat of liquid 180
Btu and 1 lb. of steam will give
up 1157.7-180 or 977.7 Btu. The
steam required per hr. will be
2,061,675 divided by 977.7 or
2108 lbs. an hr.

Average water temperature will be
117.5° and steam temperature 228°
at 5.3 psig, a difference of
110.5°. Since there is no positive
circulation heat transfer through
the tubes will be about 100 Btu
per sq. ft. of surface per degree
temperature difference per hr., or
a sq. ft. of surface will transfer,
under the conditions given, 100 x
110.5 or 11,050 Btu an hr. To
transfer the heat needed will require 2,061,675 divided by 11,050
or 187 sq. ft. of surface. Using
1 in. pipe, which takes 2.9 ft.
to give a sq. ft. of external
surface, the length needed would
be 2.9 x 187 or 542.3 ft.

Heater Horsepower: What does the
horsepower of a feedwater heater
mean?

Really not much of anything. A
heater horsepower is based on ability to heat 30 lbs. of water an
hour to within 2 or 3° of the
temperature of steam supplied, and
so has an apparent relation to
boiler horsepower. Basic temperature used by makers, however, varies
from 40 to 60°F., so that heater
horsepower really has no standard
meaning. For instance, a 750-hp.
heater on a 40° basis has the same
heating capacity as an 825-hp.
heater on a 60° basis.

Rating should be by lbs. of water
heated per hr. from initial feedwater temperature to within say 3°
of steam temperature.

Open Heater Use: Can an open
heater be used with an exhaust
steam heating system?

You can use an open heater with
any kind of heating system. In
one example it was used in connection with a vacuum heating system,
on which a vacuum of 10 ins. was
maintained constantly, with all
the way from 1 lb. vacuum to 3 lbs.
back pressure on the main supply
close to the heater. Live steam
was used when no exhaust was
available and, in case of shut
down of the vacuum pump for any
reason, straight back pressure
was used for the time being.

Open Heater Temperature: What
temperature can be reasonably expected of discharge from an open
heater with proper size piping at
1 lb. pressure?

This depends on the amount of
steam available, and the amount of
water which it is expected to
heat. If those values are properly proportioned or the heater is
being worked at its rated capacity, however, it is possible to
obtain a temperature of about
210°F.

Heater and Pump Location: Why
should a feed pump discharge
through a closed heater rather
than drawing water from the heater?

It is preferable to take the water
from the heater to the feed pump,
but in order to do this, it would
be necessary to set the pump and
heater so that water flows to the
pump by gravity, as you could not
lift the hot water by suction. It
would "Steam bind" and pump the
minute any suction was put on the
inlet chamber to the pump. Of
course, if it is not possible to
set the pump below the heater, then
it will be necessary to pump
through the heater, but the other
arrangement is preferred. (The
pressure from the pump on the water
will keep it from flashing into
steam.)

Water Heating Capacity: How many gal. of water will a coil of 100 lineal ft. of 1-in. copper or brass tubing heat per hr. when surrounded by a temperature of 500°F., the temperature of water entering the coil 50°F. and discharged at 150°F.? The tubing is to be of sufficient thickness to withstand a water pressure of 70 lbs. per sq. in. Is there any rule or table for computing the number of gal. of water that copper, brass, or iron pipe will heat to a required temperature when the size and other temperatures are known?

The method of finding the gal. per hr. that can be heated under these conditions is to multiply the area in sq. ft. of the exterior of a heating coil by the average temperature difference between the water and the gases, and by a constant. Divide this by 8.33, the lbs. of water per gal., and by the temperature rise of the water.

For 1-in. pipe, the length of 2.9 ft. will give a sq. ft. of external surface, so that 100 ft. will give about 34 sq. ft. The average temperature of the water is 100°F., and the difference between this and 500 will give 400° temperature difference.

The constant for copper or brass pipe would be about 250, and the temperature rise of the water is 100°F. Using these valves it figures to be about 4000 gal. of water an hr. that is heated by 100 ft. of copper pipe.

The thickness of the pipe makes very little difference because the chief resistance to the conduction of heat is at the surface of the pipe. For wrought iron or steel pipe the constant will be practically 200.

Heater Leakage: How can leaks in a closed heater be detected?

The most apparent indication is an increased amount of condensation. This might be so gradual, however, that you would not notice it for some time. A good plan is to test the heater occasionally by raising the water pressure in it and keeping the steam out. Any water then coming from the drips will come through leaks in the heater.

Heater Overflow: What prevents the water from backing up the exhaust pipe into the engine when an open heater is used?

The overflow valve in the heater prevents water in the heater from rising to the level of the exhaust pipe flange. There is, or should be, a drip valve in the lowest part of the exhaust pipe that should be open while starting the engine.

Heater Floods: An open heater or pan type, with water level regulator, sometimes does not work, and the heater fills with water that blows out of the exhaust head with the steam. When the heater overflows a gauge on the exhaust pipe between engine and heater shows 15 lb. pressure. Is there danger of blowing out a cylinder head or otherwise injuring the engine?

The trouble may be in the water regulator, which does not close as it should when the water has reached a certain height. On a well-designed open heater, means should be provided for letting excess water out of the heater before it can run over into the exhaust pipe.

There is always a chance for a chip or some other substance passing in with the water, and lodging under the regulating valve and preventing its closing entirely. The pressure of 15 lbs. on the heater (and consequently on the engine as back pressure), is caused by the water passing up through the exhaust pipe in intermittent slugs of steam and water, in a manner similar to the discharge of water from an air lift well.

There is no actual danger to the engine as long as it is in operation under a full and steady load, for the passage of the exhaust steam will always tend to keep the engine free from water. If the load is light or the steam pressure excessively high, the steam is likely to expand down below the atmospheric line, where as soon as the exhaust valve opens, there is danger of water from the heater being drawn up into the cylinder with disastrous results.

By all means remove this danger at the earliest possible moment. A backpressure of 15 lbs. on an engine is not conducive to economical operation, to say nothing of the danger should the load fall off at a time when the heater is full and overflowing.

Heater Tube Corrosion: Untreated water from city mains is used in a hot water heater which has brass tubes for steam coils. Water is heated to 150° and rapid corrosion takes place. How can it be prevented?

It may be due to several causes, and analysis of the water is necessary to find what it carries in solution. If it is magnesium chloride or sulphate, sodium carbonate is the cure. If free acids, an alkalide should be added. If carbonic acid or oxygen, preheating the water should help, and possibly the addition of a little lime or caustic soda. Grease or organic matter will require iron alum as a coagulant or sodium carbonate followed by filtering.

Sometimes galvanic action is set up

between the iron shell and brass tubes, which may be overcome by fastening zinc plates to the tubes. A thick coating of red lead and linseed oil or of graphite paint applied when the tubes are quite dry may stop the trouble.

Test for Heater Leakage: What is the proper way to test the coils of a closed heater for leakage?

Shut off steam, open the drain cock, and admit the cold water under twice the pressure formerly used. If leakage occurs water will be forced into the steam space, and may be collected from the drain. To find the point of leakage remove the heads, paint the pipes with a heavy soap solution and admit water pressure to the coils. Bubbles will indicate the location of the trouble.

HANDLING

Automatic Injector: What is an automatic injector?

The automatic injector has a movable part that, as the steam jet gets into action and creates a suction between the forcing tube and lifting tube, is drawn up and closes the outlet to the overflow. Until this suction is formed by the action of the steam and water jet the movable piece will rest, so that water will escape from the overflow. In case the steam into the boiler breaks for any reason the movable piece will drop back, and water will again escape from the overflow until the jet is established. The advantage is that the injector after breaking will restart of its own accord, instead of having to be shut off and restarted by hand.

Injector Action: Explain the theory of how an exhaust steam injector pumps water into a boiler against 80 lbs. pressure?

In order to use exhaust steam,

the steam nozzle of an exhaust steam injector is enlarged and a central spindle added, which, when the pressure of the boiler into which the feedwater is delivered exceeds 75 lbs., is made hollow so as to admit a supplementary jet of live steam from the boiler. The feedwater should flow to the injector and its temperature should be as low as possible, under no conditions above 92°. With the supplementary jet and cold feedwater the injector will feed against pressures up to 150 lbs. and without the supplementary jet against 90 lbs. Increasing the temperature of the feedwater somewhat diminishes these pressures.

Feed Valves: How should the stop valves on a feedwater line be placed?

The valves on the feed line to a boiler should be placed so that they close against the flow of water to the boiler. When the flow ceases the valve tends to close from the boiler pressure. In case the disc comes loose from the stem, the flow will hold it open so long as the feed is operating, and the pressure will close it when the feed stops.

Check Valve Fails: If the disc of the feed line check valves was lost, could the boiler still be operated?

If the disc of the check valve becomes loose the boiler can still be operated, provided that the pump valves are tight, because the discharge valves of the pump will prevent backing up of water from the boiler. Whenever the check valve is inoperative, however, or if it becomes necessary to shut down the pump or injector the stop valve in the feed line should be closed, and care taken to see that this is opened again when the pump or injector is started.

Feed Pipe Vibration: What changes would you make to stop the vibration and pounding in the feed line to the boiler?

Vibration and pounding in the feed line is caused by the inertia of the water that tends to continue in motion after the impelling force has ceased to act. Its energy of motion is then expended in giving a blow to the containing pipe. Any means that induces an uninterrupted flow or brings the water gradually to rest prevents water hammer. In a duplex pump the first result can be obtained by adjusting the steam valves, so that one piston begins its stroke before the other one has stopped.

The second result, gradual checking of the flow, can be secured by placing an air chamber on the discharge pipe near the end of the pump, or by regulating the pump cushion in such a manner as to bring the piston gradually to rest rather than holding its speed to the end of the stroke and then stopping abruptly.

The noise accompanying water hammer in the feed line can often be eliminated by making the pipe more secure at its point of support.

Stopped Feed Line: A 60-ft. feed line from closed heater to boiler fills with sediment from water that is apparently soft and clear, being taken from a pond fed by a large brook. The soft deposit gathers in 8 months so that the line must be cleared out. Is there a remedy?

Clear water may not be soft, and the sludge is probably precipitate thrown down after the water is heated.

One remedy is to use an open heater with a large settling action to catch the sludge in the heater. Heating the water to 180 or 200° and allowing it to pass slowly through the filtering bed of an open heater will separate most of the precipitate, in addition to any organic matter that may be in the water.

Blowoff Feeding: A 30-hp. boiler with the feedwater pipe enters the front end, just above the top row of tubes. Can water be entered through the blow-off pipe to get rid of a red sediment that forms on the tubes?

There is considerable difference of opinion as to the advisability of feeding the boiler through the blowoff. It is done in many plants, and if there is not much formation of scale it probably is not harmful. If there is formation of scale it tends to produce a current that carries the sediment away from the blowoff and toward the front end of the boiler. There it lodges over the fire, and is more likely to produce trouble than if carried toward the back end of the boiler.

Also, putting the water in through the blowoff gives no chance for heating the water inside the boiler before it is discharged against the tubes.

A better plan is to carry your feed pipe from its present ending inside the boiler, back well to the rear end, and then turn it down, either between the rows of tubes in the center, or carry it across to one side and then down along the shell of the boiler, discharging near the bottom. In this way, by passage through the feed pipe inside the boiler, the water becomes heated practically to boiler temperature before it is discharged to mingle with the circulation of the boiler. You can bring the discharge to any point where you prefer to have the sediment deposited, but naturally this would be near the blowoff.

Part 6: Steam

Wetness of Steam: Does exhaust steam at 5 lbs. pressure contain more water than live steam of the same pressure?

Exhaust steam usually contains more water or moisture than live steam of the same pressure. Starting with high pressure steam, which is practically dry and has no moisture in it, during the expansion through the engine some of this is condensed, and its heat of evaporation converted into work. This condensation is the exhaust steam.

The amount of such condensation depends on the amount of the expansion, that is, the pressure range between the initial pressure and the exhaust pressure. There is also some condensation due to heat absorption by the cylinder walls and pistons.

On the other hand live steam at 5 lbs. pressure is usually obtained by passing through a reducing valve from a higher pressure. In this very process of passing through the reducing valve, or throttling, some of the moisture that might be in the live steam is evaporated. The steam's temperature is approximately the same as that of the high pressure steam making it now at 5 lbs. superheated steam.

Combining Steam at Different Pressures: Mixing superheated steam with saturated steam is, of course, unusual. A given plant, though, uses two separate boiler plants: No. 1 being equipped with superheaters and delivering steam at 200 lbs. pressure and 100° superheat; No. 2 boiler plant generates saturated steam at 145 lbs. and the boilers are not safe for higher pressures. No. 1 plant supplies steam to turbines direct, and also supplies steam at 145 lbs. pressure through a reducing valve to the No. 2 heater that supplies steam to several engines.

Assuming No. 2 does not generate enough steam to supply the 145-lb. demand, superheated steam is admitted automatically, and flow continues until the pressure builds up. This procedure would normally be gradual, but there is an instantaneous demand at times for about 500 boiler hp. for the digesters. Such a condition would call for a sudden pull on the superheated steam if No. 2 happened to be running under light load at the time. One theory has it that superheated steam will combine with saturated steam without disturbance, drying out and raising the temperature of the latter. The possibility of hammering and internal disturbances, however, is disquieting. Does this procedure involve complications?

The trouble is due to collection of condensed steam in the bleeder line because this is doubtlessly frequently deadened, at which time it is inactive. Introduce a steam trap or traps at all low points in this line so that there will be no condensate in the pipe when the steam from the superheat is turned into the line.

Superheat: Why is superheat used?

Moisture in the steam supplied to a prime mover, whether a steam cylinder or a turbine, lowers economy of operation. Even if steam is dry as it enters the prime mover some moisture of condensation will be produced by contact with the containing walls and by extracting heat to do the work. When superheated the steam can give up heat without moisture resulting, hence the increase in economy by reducing cylinder condensation in engines and friction in turbines. Also, blade erosion is avoided in turbines. Theoretical economy from a heat standpoint is small, but actual steam consumption is increased roughly 1%

for each 12 to 15° of superheat. American practice in superheat runs to temperatures of 450 to 740°. Some European plants use as high as 850°F.

Superheat Calculations: How is the degree of superheated steam figured? For example, when the steam pressure is 140 lbs. gauge and it is superheated to 470°F., what is the superheat?

To find out the amount of superheat in steam at 140 psig, look at the steam tables to find the temperature corresponding to this pressure, which must first be changed to its corresponding absolute pressure by adding 15 lbs., or approximate atmospheric pressure, to the gauge pressure. This we find to be 140 plus 15 or 155 psia. Consulting the steam tables we find the temperature corresponding to dry and saturated steam at 155 psia to be 361°. Since the steam in the problem actually has a temperature of 470°F., the amount of superheat is 470 minus 361, or 109°.

Part 7: Steam Engines

VALVES

Valve Leakage: How may steam valves on a Corliss engine be tested for leakage?

Steam valves on a Corliss engine may be tested for tightness by taking off the bonnet on the valve and admitting steam into the cylinder through the opposite steam valve.

Piston vs. Slide Valve: What advantage has a piston valve over the slide valve? Ports on a piston valve may be short and direct, thus reducing clearance, volume, and surface. Operation is prompt, thus avoiding wire drawing, and the valve is balanced, thus reducing friction.

Eccentric Rod Length: What is the effect of lengthening the eccentric rod of a slide valve engine?

This will move the valve toward the head end of the cylinder, giving more steam lap and less exhaust lap on the head end. This results in later admission, earlier cutoff, earlier release, and later compression. For the crank end the effects would be reversed.

Negative Lap: When and why is negative lap used on Corliss engines?

Nearly all single eccentric Corliss valve gears and some double eccentric gears are of the so-called half-stroke type, in which the steam valves usually have positive lap, with wristplate central. Maximum cutoff about four-tenths stroke. To increase the range of cutoff the steam valves are sometimes given negative lap, the eccentric being retarded in proportion.

With the long range gear having a maximum cut off of three-fourths stroke, the steam levers travel right with the rocker arm so that with the latter in central position, the steam valves are half open. A steam wristplate need not be used. With this type of gear it is necessary to have the steam hooks trip each stroke, regardless of the position of the governor. The exhaust valves always have some positive lap, and are treated the same in half stroke gears as in long range.

Positive Cutoff: What is meant by positive cutoff?

Engine valve gears known as positive cutoff are those which do not release the valve from the gear at the point of cutoff. The term "positive cutoff" is regularly applied to the "four-valve" engine or so-called non-releasing Corliss engine.

Vertical Engine Lead: How is the amount of extra lead on the lower end of a large vertical engine, to take care of the weight of reciprocating parts, determined?

It is usually a matter of trial. The object is to bring the reciprocating parts to rest without strain on the crank pin, which may be done partly by increased compression and partly by earlier lead. As a rule compression and lead should be as small as is consistent with quiet running.

Fly Ball Governor: What causes a fly ball governor to race?

Several things may make the engine "hunt" or "race" for several seconds when the load goes off. Excessive friction of the cutoff device, caused by accidental strain of some of the rods or connections, will tend to retard the governing action of the governor. These parts must be absolutely free from strain of any kind.

BOILER OPERATION

Care of Steel Boilers When in Service

GETTING READY TO START

OPEN SYSTEM--An open-type hot water heating system is provided with an expansion tank located at highest point of system. An overflow pipe is connected to the tank to prevent excess pressure. An altitude gauge indicates height of water in expansion tank.

1. When filling boiler notice movement of altitude gauge hand and ascertain that black and red hands are together just before water overflows from expansion tank.

2. Admit sufficient water to make sure that overflow pipe is clear.

3. Do not run overflow outside.

CLOSED SYSTEM--A closed type hot water heating system is provided with a reducing valve and a relief valve to prevent excess pressure, and also with an altitude gauge or a pressure gauge to indicate amount of water in system.

1. Black and red hands of altitude gauge should be together when water stops flowing into system.

2. Relief valve should be tested by pulling hand lever, if there is one, to determine that it works properly and that disc is not corroded to seat.

STEAM HEATING SYSTEM--A steam heating system is provided with a safety valve to prevent excess pressure; a steam pressure gauge to indicate pressure within boiler; and a gauge glass and gauge cocks to show level of water in boiler.

1. Observe rise of water level in gauge glass and open gauge glass drain cock and gauge cocks as boiler fills with water.

2. Test the accuracy of gauge glass by using try or pet cocks on water column.

3. If water does not flow out freely obstructed cock should be cleaned immediately.

4. It is of great importance that water level indicators operate properly, for low water may ruin boiler or cause a boiler explosion.

5. Valves at top and bottom of water gauge glass should always be open.

6. Safety valve should be tested by pulling hand lever to be certain that it works properly and that disc of valve is not corroded to seat. (Always pull when there is at least 70% pressure so as not to damage the valve.)

7 Make sure that steam gauge cock is always open. (Have two gauges because the code requires that the boiler be shut down when the gauge is broken.)

STARTING THE BOILER

1. Make certain that valves in piping to heating system are open before starting fire.

2. Dampers must be opened and firebox thoroughly vented.

3. Breeching must fit tightly into chimney and should not extend into stack beyond thickness of wall.

4. If solid fuel is used, start fire as directed under "Firing Instructions," but do not use gasoline or other highly inflammable substance.

5. If liquid or gas fuel is used, follow instructions of manufacturer of burner that should be posted in boiler-room. Do not turn on fuel supply at any time until firebox has been thoroughly ventilated or until sure a lighted torch or ignition system will ignite fuel immediately at low fire.

6. Increase temperature gradually—watch water level carefully. If it appears necessary to admit more water, do so, but determine reason why more water is needed (loss of water from system indicates there is a leak). Do not confine investigation to boiler but examine entire system.

7. If at any time burner is shut down, dampers should be kept open after fuel is shut off to free firebox of inflammable vapors which otherwise may accumulate. (Vent the fire box.)

8. Oil and gas fuel are automatically controlled, but do not leave boiler room after starting fire until it has been determined that burner and automatic regulators are in working order. Do not depend entirely upon automatic controls but make a practice of visiting boiler room periodically to see that everything is working properly.

ROUTINE OPERATION

1. Do not carry more steam pressure than is absolutely required to heat building.

2. Do not try to operate multiple boiler heating system with one boiler when weather is cold and two boilers are required.

3. If only one boiler is needed during part of day, do not close valves of idle boiler but allow it to remain warm and connected to system, so that it can be fired up at any time—operation of system is simplified in that way.

4. If steam and return valves have been closed for any reason during the time one boiler or battery was out of service, those valves should not be opened again until pressure of entire system is equalized, preferably by reducing pressure to zero on boiler that is in service.

5. If it is impossible to reduce pressure idle boiler should be fired up and pressure raised to an amount exactly equal to that of boiler in service before valves are opened.

6. Return valves should be opened first and steam valve immediately thereafter. (Open slowly.)

7. Operation of boiler respecting both pressure and water level should be observed carefully for some time after boilers are connected together.

8. Water level in gauge glass of steam boiler should not vary greatly.

9. If water disappears from the gauge glass as temperature is increased each morning, and again appears at its usual level when fire is low, a thorough investigation should be made.

10. Hot water must not be drawn from heating boiler for any purpose whatsoever; to do so may result in damage to boiler.

11. Water returning from system is ample for boiler if system is tight. Need of additional water to greater extent than one inch in gauge glass of steam system or 1 foot, according to altitude gauge of water system, per month, is indication that water is being lost from system.

12. Excess amounts of make-up water results in formation of scale within boiler.

13. Water when needed should be added slowly and at time when load is lightest.

14. Determine from experience the pressure or temperature of system that affords best results with changing weather conditions, and maintain proper pressure or temperature as nearly constant as possible with uniform fire covering entire grate.

LOW WATER

1. If at any time water is not visible in gauge glass or its height does not register on altitude gauge, put out the fire at once. Allow boiler to cool until crown of firebox is comfortable to hand before adding water. Inspect if you have any doubts.

2. If no burns are found after increasing amount of water in boiler to normal level, determine cause of low water condition and correct difficulty before starting fire.

SAFETY APPLIANCES

1. Gauge cocks, gauge glass and safety valve of steam system should be operated by hand each day.

2. Water relief valve of closed hot water system must be piped to discharge in safe place so that it may be operated by hand at least once each month. No valve can be connected after the valve discharge or between the boiler and relief valve. The discharge line diameter must not be reduced, and must not be in an area where the temperature will freeze the discharge water. It must discharge in a safe place.

3. Overflow from expansion tank of open hot water system must discharge in safe place, and sufficient water should be added to system each week to ascertain that overflow pipe is clear and that altitude gauge registers correctly.

4. Safety valve of steam heating boiler should operate freely before steam gauge registers more than 15 lbs. If it does not, pressure should be reduced to zero, and boiler should be cooled as promptly as possible so that safety valve or pressure gauge may be adjusted or repaired. No valve should be connected between the boiler and the safety valve or between the safety valve and discharge pipe. The discharge pipe must be reduced. It must discharge to a safe place. It must be mounted without damage to its body by a tool.

BLOWING OFF

1. Under normal conditions a heating boiler does not need blowing off, emptying or cleaning during the heating season if chemical is added.

2. If it becomes necessary to empty boiler for any reason, be sure that fire is out and that firebox is comfortable to hand.

When new boiler is placed in service for first time, it should be cleaned in the following manner:

(a) Fill boiler until water reaches top of gauge glass.

(b) Add soda ash or caustic soda at rate of 5 lbs. per 1000 sq. ft. of connected radiation.

(c) Dissolve chemicals thoroughly before introducing into boiler.

(d) Install a vent pipe to convenient drain.

(e) Start light fire sufficient to boil water without producing steam pressure.

(f) Open boiler feed valve

sufficiently to allow steady trickle to run from overflow pipe.

(g) Boil slowly for 3 or 4 hours.

(h) Let fire burn out and dump remaining live coals into ashpit where they can be quenched with hose.

(i) Remove handhole covers or washout plugs, and hose out boiler thoroughly.

(j) Fill to normal water line.

(k) If water in gauge glass is not clear, repeat procedure, using stronger solution of soda ash or caustic soda, and boil for longer time. Remove drain, install safety valve.

(l) Aggravated cases of dirty boilers are sometimes better corrected by blowing off under pressure. This should preferably be done by someone familiar with that method.

CLEANING

1. If possible the tubes should be cleaned each day and at least once each week during heating season.

2. Cleaning should be done when load is lightest.

3. Soot prevents heat from passing in water surrounding tubes, and when moist will cause serious corrosion.

4. When soot accumulates in tubes more coal must be burned to offset reduced efficiency, and more ash must be removed. The resulting work is greater than that involved in keeping tubes clean.

5. Remove fly ash from smokebox.

Firing Instructions

STARTING FIRE

1. Check the water level and gauge glass operation. Cover entire grate with kindling varying in depth from 5" at front to 10" or 12" at bridge wall.

2. After kindling has been lighted and is burning spread a 3" or 4" layer of lump coal over entire grate of Direct Draft Boilers and in front of arch of Smokeless Boilers.

3. When this coal is well ignited spread on another layer.

4. When this is thoroughly ignited push most of fire back against bridge wall, leaving a thin layer on front of grate to ignite new charge.

5. Throw next charge into front portion of firebox.

ROUTINE FIRING (COKING METHOD)

1. Large lumps of coal should be brought to 6" or smaller before firing.

2. When firing damper in breeching should always be wide open; cold air check, ashpit and draft doors should be closed.

3. To attend boiler first shake grates until live ashes fall into ashpit.

4. Push live coal to rear, leaving a thin layer over the front portion of grate, and recharge green fuel in front of arch of Smokeless Boilers or on front two-thirds of grate of Direct Draft Boilers.

5. A heavy fire is more economical under almost all conditions.

6. Never pile fuel tight against arch. Always leave from 3" to 6" clear space.

7. Never cover entire fuel bed with green coal, not even when fire is to be blanked (this applies to Direct Draft and Smokeless Boilers). Explosions may result if these instructions are not followed.

8. Disturb the fuel bed as little as possible, but do not permit holes to develop in fire even when only a little heat is required. Holes in fire permit excess air to pass through boiler, materially reducing its efficiency.

9. If coking coal is used it may be necessary to slice fuel bed about 30 minutes after firing.

10. Do not permit fire to burn too low before recharging. High temperatures are necessary to obtain smokeless combustion.

11. Enough coked fuel should remain each time to cover entire grate with a thin layer of live coke, and a thick layer just ahead of bridge wall.

12. To clean fire push live coals to rear. Shake rear grates. Then pull live coals to front, and shake rear grates.

13. Do not leave ashpit door closed after dropping live coals or clinkers into ashpit.

14. If necessary allow ashes to accumulate on grate to retard fire in mild weather.

15. When boiler operates at high capacity it may be necessary to reduce fuel depth.

Care of Grates

1. Grates are intended to support fuel and to permit a sufficient amount of air to pass through same.

2. Air must pass through openings in grates, and there must be sufficient space beneath grates for the air to pass to rear of ashpit.

3. When ashes are permitted to accumulate beneath the grates, air cannot reach rear portion of ashpit.

4. Ashpit must always be kept clean. Leaving red hot ash in the ashpit will ruin grate support bars, grates, and the boiler base.

5. If cleaning of ashpit is neglected it is a certainty that new grate bars will be required.

6. Ash will always be found at bottom of fire with hot burning fuel directly above.

7. Do not shake all ash through to ashpit; allow a thin layer to protect grate bars from heat of fire.

8. Ash from certain grades of coal has a low melting point, and if not taken care of properly will stick to grate bars, closing up air spaces.

9. After shaking grates make sure that shaker lever is in vertical position before removing shaker handle. This will make sure the grates are not tilted.

10. It is also very important in banking a fire not to add fuel in excess of that required because if the resistance to the passage of air through the fuel bed is too great, and ashpit and draft doors are closed, heat from fuel bed may cause warpage of grates.

11. In new buildings ashpits should be placed in floors by merely inserting a wood form when boiler room floor is poured.

12. Ashpits are desirable on all large boiler installations. With them it is possible to keep ashes farther away from grates.

13. Whether ashpits are installed or not entire space below grates should be cleaned at least once each day because an accumulation of ashes will burn out or warp grates.

14. To bank fire proceed as in preparing fire for regular operation except that a greater charge of fuel should be used. Move regulator weights to opposite side of regulator. When fire is banked it will not continue to make steam.

15. In breaking up a banked fire move regulator weights to usual position. Shake grates, break fuel bed so air can get through, and let fire burn up until live coke is produced before adding green fuel.

AUXILIARY OR SECONDARY AIR

1. The use of auxiliary or secondary air is very important in boilers burning bituminous coal.

2. Air should be admitted through doors on side of Smokeless Boilers during entire period when fire is being cleaned and new fuel is being added.

3. These doors should be left open for a period after this operation. The exact time can be determined by operator after a little experimenting.

4. With Direct Draft Boilers it is advisable to leave the slides in fire door open at all times when boilers are used with bituminous coal.

5. When starting fire it may even be advisable to leave fire door slightly open.

Care of Steel Boilers When Not in Service

CLEANING

1. Clean boiler thoroughly inside and outside.

2. Scrape tubes down to metal.

3. Brush down front and rear heads, inside of firebox and crown sheet.

4. Remove all soot, scale, rust or other deposits with steel brush.

5. Clean grates.

6. Remove ashes from edges and corners of firebox.

7. Clean soot and ashes from behind bridge wall.

8. Clean ashpit thoroughly.

9. Clean out smokehood and breeching.

10. Clean outside of boiler where uncovered.

WASHING

1. Drain boiler.

2. Remove handhole and handhole covers and washout plugs.

3. Wash boiler out thoroughly with hose, using sufficient water pressure to remove all sediment and scale.

4. Place hose in openings to flush out water legs on all four sides.

5. Leave off all manhole and handhole covers and washout plugs so air may circulate through boiler.

6. Keep feed valve closed so no water can enter boiler.

DRYING

1. Dry boiler by placing small lighted single burner kerosene stove or single burner gas plate inside of firebox.

2. Let it burn until boiler is dry.

3. Do not dry boiler by building fire upon grate; there is danger of damaging boiler even though very light materials are used.

OILING AND PAINTING

1. Swab all tube surfaces with a rag soaked in mineral oil.

2. Apply coat of boiler paint or mineral oil to exposed outside surface of boiler.

CAUTION

1. If boiler remains empty when not in service, extreme caution must be taken to see that fire is not started.

2. Place a seal or a caution sign on fire door.

3. If hazard is too great replace manhole, handhole covers, and washout plugs.

4. Fill boiler with water until heat is required.

5. Then drain all water, refill and place in service.

6. If boiler has been without water during idle period replace manhole and handhole covers and washout plugs; then fill with water to correct level.

7. A steel boiler which remains filled with water and coated with soot when not in service deteriorates rapdily.

8. Refuse should not be burned in boiler during the summer unless a responsible individual determines that boiler is filled with water.

9. The burning of wood and paper sometimes results in a higher rate of burning than would be experienced during the heating season.

10. Such material should be fired in small quantities if it must be consumed in heating boiler.

11. A boiler used for such service is usually not properly cleaned before the heating season begins.

12. Keep boiler room dry to avoid rusting in summer.

13. Leave flue doors, fire doors and ashpit doors wide open during the entire period while boiler is not in use; this will keep the flues dry.

INSPECTION

1. At end of heating season clean and inspect boiler, grates and accessories. Clean gauge glass or replace. Make sure the gauge connections are not plugged with dirt.

2. Place in good working condition for next heating season.

3. Repair parts are more readily obtained during summer season and installation charges are frequently less than in fall.

4. Do not neglect boiler--clean as directed above.

5. More damage can occur in one month when boiler is idle than in entire heating season if care is not taken to properly prepare boiler for idle period.

Automatically Fired Boilers

STARTING NEW BOILER

Before starting the burner make certain that there is water in the boiler. Water level should be carried approximately at the middle of water gauge glass for a steam boiler. The required amount of water in a hot water heating system may be determined from indication of the hand on altitude of the boiler. Water will appear at the air valves of every radiator when the system is full of water.

A feedwater line with valve will be connected into the return main near the boiler for the purpose of adding water. This valve must be kept closed. The boiler blowoff valve and any drain valves to the heating system must also be closed.

When provided the valves in supply and return mains to the heating system must be kept wide open when the heating plant is operated. If these are closed for summer time hot water service both valves must be closed tight.

WATER GAUGE

The valves at top and bottom of the water gauge glass, provided with steam boilers, must be kept open so that the true boiler water level will be shown in the gauge glass.

The try-cocks or gauge cocks at the side of the water glass are for the purpose of checking the indication of the water gauge. Open these occasionally when there is pressure on the boiler.

With the drain cock at bottom of gauge glass open and with steam pressure on the boiler the valves should be occasionally closed and opened alternately to blow them clean of any sediment. Do this two or three times each heating season. Keep the stuffing box on the valves tightened to prevent air leakage.

Put a spoonful of muriatic acid (hydrochloric) in a cup of water. With pressure on the boiler close both water gauge valves and open drain cock at bottom of glass, allowing water to flow out. Open top valve allowing steam to flow through glass and out of drain cock; allow the steam to flow until glass is thoroughly heated. Then close the top valve and place cup of solution so that the drain cock is submerged causing the solution to be drawn into the water glass. By keeping the drain cock in the solution and alternately opening and closing the top gauge valve the glass will be thoroughly cleaned.

Be sure when the cleaning is completed that both top and bottom gauge valves are again opened. Safety valve should be tested once a month by pulling the lever when the boiler has 75% pressure or at the time when the pressure throw shuts off.

Automatically fired boilers should be provided with water level controls and these must be tested at least once a week as follows: With the thermostat set up so the burner will operate and with the water level normal, open the blowoff valve to let out water. The low water cut-off control should stop the burner when the water level is within 1/4 inch from the bottom of the glass.

Some low water devices have a valve that is spring loaded to shut off the water drained from the device, or else they have a gate valve or a ball valve with the opening the diameter of the drain. Keep running water on it until it is free of rust and mud. Collected mud causes failure causing the boiler to distruct on a low water condition in that it will fail to shut off the fire in the boiler.

Close the blow-off valve and open the feedwater valve. When the water level is raised back to 3/4 inches in the glass, the cut-off control should again allow the burner to operate. Where an automatic water feeder is used it should stop feeding water to the boiler when the level is 1 inch above the bottom of gauge glass. The drain valve should be opened at regular intervals, and always directly after cleaning the boiler to clear the chamber of any sediment.

Steam heating boilers are equipped with a pressure limit control switch and hot water boilers with a temperature control switch, to govern the operating range of pressure or temperature. The pressure or temperature at the boiler should be no higher than is required to heat the building in coldest weather.

DRAFT ADJUSTER

The draft adjuster is designed to maintain a uniform draft on the boiler. Regulate the damper plate to swing freely. For strong chimney drafts the damper should be set to swing nearly wide open and for low drafts it should be kept practically closed. Leakage of furnace gases into the boiler room is usually an indication of low draft.

CHIMNEY

The smokepipe connecting the boiler with chimney must have tight fitting joints. All cracks should be sealed with boiler putty or cement. Separately fired water heaters, incinerators or fireplaces must not be connected to the same chimney flue with the heating boiler.

CHECKING YOUR PLANT

A high pressure plant tender is expected to be able to check the efficiency and performance of the individual units in the plant, and to be able to make recommendations and suggestions for increasing performance and capacity.

Assume that you have been asked to check an average pulverized fuel fired steam generator. Write out the methods you would follow in checking the following units, and what suggestions you might make to produce more steam.

    1. The Pulverizer
    2. Burners
    3. The Furnace
    4. The Boiler
    5. The Superheater
    6. The Economizer
    7. The Air Heater
    8. The Induced Draft Fan
    9. The Forced Draft Fan
    10. Boiler water tests.

When you have completed your answers, then check the following answers, which have been prepared by the Foster Wheeler Corporation.

ANSWERS

1. Pulverizer: Make a careful investigation of pulverized coal fineness, selecting an average sample in accordance with the accepted methods of testing. Poor burner performance, ex-

cessive slagging and high combustible loss, all of which may reduce maximum steaming output, can be due to coarse grinding.

Reclassification of ball charge may be necessary. Wedge bars should be raised to reestablish their original lift by fitting shims between the sides of the wedge bars and liners.

2. Burners: Examine and repair burner throats, if necessary. With circular turbulent type burners, poor flame shape and flame impingement on side walls, as well as excessive combustible loss and localized slagging, may result from burner throats having deteriorated from original dimensions. Improved burner performance means added capacity and efficiency.

Examine and, if necessary, repair burner coal nozzles. On many types of burners poor flame travel and lack of adequate mixing of coal and air may result from over-extending the life of coal nozzles. Nozzles exposed to radiant heat, or where ignition has been maintained too close to the burner, may grow to such an extent that velocity of coal-air mixture leaving the nozzle is much lower than was originally intended. Inability to obtain proper flame travel results.

3. Furnace: Observe disposition of flame in the furnace and distribution of coal between burners. Coal distribution may be checked by carefully timed sampling in the coal conduits. A better test is to analyze flue gas at entrance of boiler and distribution of CO_2 across width of furnace. Poor fuel distribution may result in localized slagging of boiler entrance and superheater, as well as furnace walls and ashpit, and so limit the output of unit from these causes. Instruments which have been adjusted to indicate CO_2 at boiler outlet may, because of serious air infiltration through the setting, result in operation with entirely too high a CO_2 in the furnace. This can only be determined by sampling flue gases through a water-cooled sampling tube at entrance of boiler.

Examine furnace setting, particularly ashpit and seals between ashpit and furnace, for air leakage. Poor combustion may result from burners getting an inadequate supply of burner air, and too great a percentage of leakage air. While some of this leakage air enters into the combustion process, nevertheless, for a given CO_2 leaving the furnace, best burner performance will be obtained with maximum air entering the furnace through burners and a minimum as leakage. Furthermore, all air that enters the furnace as leakage is that much less air through the air heater. This results in a corresponding decrease in overall unit efficiency.

4. Boiler: Check boiler exit gas temperature by a careful traverse with a thermo-couple to obtain a true average temperature. Compare this with a guarantee. High exit gas temperature will enable greater output to be obtained. A carefully made traverse will show up leaky baffles, setting leaks, poor fuel distribution, and whether or not the surface is dirty. Tubes containing internal scale should be turbined and services of a competent feedwater specialist retained to examine the feedwater treating problem and recommend means for scale elimination. Presence of external scale and soot may indicate repairs to soot blowers, installation of additional soot blowers, or increase in soot blower operating pressure. If necessary, install access doors so that inaccessible parts of unit where slag accumulates may be hand-lanced during operation.

Check draft loss across boiler, and if excessive check draft loss across the individual passes. Particularly in older designs, some restriction may be present which can be removed without materially affecting efficiency, and result in ability to carry an appreciably higher load, particularly if the draft has been the limiting factor. In other cases, baffling may be opened up with some reduction in efficiency resulting, but usually the lowered efficiency is not enough to offset the gain in maximum output.

Check drop in CO_2 from superheater entrance to boiler outlet to ascertain amount of air infiltration, then seal up the setting as well as possible and check again. Keep testing until drop in CO_2 is not more than 1%; preferably 0.5%. Air leakage into setting increases draft loss through the boiler and heat recovery equipment, and is a detriment in every respect.

5. Superheater: Determine final steam temperature by a thermometer in the thermometer well at superheater outlet header, and check pressure drop through the superheater. Compare with guarantee. Pressure drop through superheater may be limiting output of unit. After years of operation, some internal fouling due to carryover, may have reduced final steam temperature as well as limited output due to increasing the pressure drop. Special instructions for boiling-out and flushing-out a superheater to reduce pressure drop are available. It is possible that additional superheater surface can be installed to increase the final steam temperature, and at the same time reduce pressure drop, thus removing this as a limiting factor on output of the unit.

Where the external surface of the superheater has become badly scaled, or covered with adherent slag coatings, sandblasting may be necessary in order to return the elements to their original condition.

6. Economizer: Check water rise through economizer and compare with guarantee. A low temperature rise will indicate necessity for cleaning economizer internally, externally, or perhaps both. Check the CO_2 drop across economizer, then candle the casing and seal all air leaks until CO_2 drop is negligible. The closer the induced draft fan or stack is approached, the greater the necessity for sealing leaks, due to greater suction prevailing the ducts and economizer itself.

7. Air Heater: Check air heater performance. This requires that careful temperature traverses be made to get the average air and gas temperatures entering and leaving the air heater. Check CO_2 drop and eliminate all possible leakage. This may require rerolling tubes, welding them, or replacing sections of plate type heaters. It may be desirable to wash the surfaces on the gas side to remove deposits that have accumulated over a period of years. In tubular air heaters, if too great a number of leaky tubes have been plugged, they should be replaced with new tubes in order to reduce draft loss.

8. Induced Draft Fan: Check performance of fan by calculating weight of gas handled, measuring total head, temperature, and power requirement. Comparison of CO_2 at the induced draft fan outlet with that entering boiler will show whether or not still further attempts to reduce air leakage are desirable. Gas temperatures leaving the induced draft fan will indicate whether or not temperature traverses of gas leaving the air heater have been carefully made. The fan manufacturer will advise whether size of fan rotor may be increased, if this is a factor limiting output of the unit. It may be that repairs to the rotor by welding have so affected fan efficiency that an appreciable gain in fan capacity may result from complete rebuilding of the fan rotor.

9. Forced Draft Fan: Check performance of forced draft fan. Efficacy of forced draft fan may have been reduced by having to handle air at considerably higher temperature than was anticipated. Some improvement may result from making air

available to the fan at lower temperature. Air heater recirculating dampers may leak or may affect flow of air to fan inlet. Since such recirculating connections were installed to prevent air heater corrosion at low loads, they are not necessary when sustained high loads must be carried and may be conveniently removed for the duration. If forced draft fan is the bottleneck a check with fan manufacturers will determine if a larger diameter rotor may be installed. The seals between the rotor and fan inlet should be examined to see that they are in proper condition. The possibility of reducing resistance on air side of system should be investigated. It is possible that some air may be by-passed around air heater and pressure loss through air heater reducer, enabling an increase in forced draft fan capacity.

Burner pressure drop may be reduced by increasing area of burner throat or by opening wider the vanes used to give secondary air a circular motion. If an unreasonable pressure drop exists in air ducts from the forced draft fan to air heater, and from air heater to burners, some gain may result from installation of guide vanes wherever turns occur in ductwork.

Boiler Inspection

Inspection of a boiler to assure conformance with the local or state boiler code, and for general safety is an important function of this position. The following directions for complete inspection of an installed boiler are taken from the New York State Boiler Code and would satisfy the requirements in most states and cities. Study this material carefully as it covers a wide range of materials from which questions on the examination may be drawn.

Rules for Inspection of Installed Boilers

General Instructions. First remove the low water cut-off. Examine its working parts for wear and tear. It is essential to have every part of a boiler that is accessible open and properly prepared for examination, internally and externally. All boilers have openings through which an examination may be made and which for operation are closed; all such parts shall be opened whether for access to water surfaces or heat surfaces. In cooling a boiler down for inspection or repairs, the water should not be withdrawn until the setting is sufficiently cooled to avoid damage to the boiler, and when possible allowed to cool down naturally. It is not necessary, in order to comply with ordinary prudence, to remove insulation material, masonry or fixed parts of the boiler, unless defects or deterioration peculiar to certain types or inaccessible parts of boilers are suspected, and where there is moisture or vapor showing through the covering, the covering should be removed at once and a complete investigation made. Upon sufficient visible evidence or suspicion due to age or other causes, every effort shall be made to discover the true condition, even to the removal of insulating material, masonry or fixed parts of a boiler. Sometimes drilling or cutting away of parts is justifiable and necessary to positively determine this condition.

The inspector shall get as close to the parts of the boiler, both internal and external, as is practicable, in order to get the best possible vision of the surfaces, and use a good artificial light if natural light is not sufficient. Whenever attachments or apparatus require testing, the tests shall be made by a plant operative in the presence of the inspector, unless otherwise desired.

Scale, Oil, Etc. Upon entering a boiler, the inspector shall examine all surfaces of the exposed metal to observe the action caused by the use of water, oil, scale solvents, or other substances which may have intentionally or unintentionally gone in with the feed water. Any evidence of oil shall be carefully noted. The smallest amount of oil is dangerous and immediate steps shall be taken to prevent any further entrance of oil into the boiler. Oil or scale in the tubes of water-tube boilers or on plates over the fire of any boiler is particularly bad, often causing them to rupture.

Corrosion, Grooving. A given amount of corrosion along or immediately adjacent to a seam is more serious than a similar amount of corrosion in the solid plate away from the seams. Grooving along longitudinal seams is especially significant as grooving or cracks are likely to occur when the material is highly stressed. Severe corrosion is likely to occur at points where the circulation of the water is poor and such places should be examined most carefully for evidences of corrosive action.

For the purpose of estimating the effect of corrosion or other defects upon the strength of a shell, comparison shall be made with the efficiency of the longitudinal riveted joint of the same boiler, the strength of which is always less than that of the solid sheet.

All flanging shall be thoroughly inspected and particularly the flanges of circular heads are not stayed. Internal grooving in the fillet of such heads and external grooving in the outer surfaces of heads concave to pressure is very common since there is a slight movement in heads of this character which produces this kind of defect. Some types of boilers have what is known as the OG or reversed flange construction in some of their parts that may be inaccessible to the eye, but the condition shall be determined by the insertion of a mirror which at a proper angle will reflect back to the eye the condition of such a place, or any other feasible manner.

Stays. All stays, whether diagonal or through, shall be examined to note that they are in even tension. All fastened ends shall be examined to note whether cracks exist where the stays are punched or drilled for rivets or bolts, and if not found in proper tension, the inspector should recommend their proper adjustment.

Manholes and Other Openings. The manhole and other reinforcing plates, as well as nozzles or other connections flanged or screwed into a boiler, shall be examined internally as well as externally to see that they are not cracked or deformed, and wherever possible observation shall be made from inside of the boiler as to the thoroughness with which its pipe connections are made to the boiler. All openings to external attachments, such as water-column connections, openings in drypipes and openings to safety valves, shall be noted to see that they are free from obstructions.

Fire Surfaces--Bulging, Blistering, Leaks. Particular attention shall be given to the plate or tube surfaces exposed to the fire. The inspector shall observe whether any part of the boiler has become deformed during operation by bulging or blistering; the former is a distortion of the entire thickness of the plate or tube where it takes place, while the latter is a lamination or separation of the plate due to foreign material being embedded in the ingot before the plate is rolled. If bulges or blisters are of such size as would seriously weaken the plate or tube, and especially when a leakage is noted coming from those defects, the boiler shall be discontinued from service until the defective part or parts have received proper repairs. Careful observation shall be made to detect leakage from any portion of the boiler structure, particularly in the vicinity of seams and tube ends. Fire tubes sometimes blister but freely collapse, and the inspector should look through the tubes for any such defects. If they are found with a sufficient degree of distortion, they should be removed.

Lap Joints, Fire Cracks. Lap-joint boilers are apt to crack where the plates lap in the longitudinal or straight seam; if there is any evidence of leakage or other distress at this point, it shall be thoroughly investigated, and if necessary, rivets removed or the plate slotted in order to determine whether cracks exist in the seam. Any cracks noted in shell plates are usually dangerous except fire cracks that run from the edge of the plate into the rivet holes of girth seams. A limited number of such fire cracks is not usually a very serious matter.

Testing Staybolts. The inspector shall test staybolts by tapping one end of each bolt with a hammer, and when practicable a hammer or other heavy tool should be held on the opposite end to make the test more effective.

Tubes--Their Defects, Etc. (a) Tubes in horizontal fire-tube boilers de-

teriorate more rapidly at the ends toward the fire, and they should be carefully tapped with a light hammer on their outer surface to ascertain whether there has been serious reduction in thickness. The tubes of vertical tubular boilers are more susceptible to deterioration at the upper ends when exposed to the products of combustion without water protection. They should be reached as far as possible either through the handholes, if any, or inspected at the ends.

(b) The surfaces should be carefully examined to detect bulges or cracks, or any evidence of defective welds. Where there is strong draft the tubes may become thinned by erosion produced by the impingement of particles of fuel and ash, or the improper use of soot blowers. A leak from a tube frequently causes serious corrosive action on a number of tubes in its immediate vicinity.

(c) Where short tubes or nipples are employed in joining drums or benders, there is a tendency for waste products of the furnace to lodge in the junction points and such deposits are likely to cause corrosion if moisture is present. All such places should be thoroughly cleaned and examined.

Ligaments between Tube Holes. The ligaments between tube holes in the heads of all types of fire-tube boilers and in shells of water-tube boilers should be examined. If leakage is noted, it may denote a broken ligament.

Steam Pockets. Steam pockets on fire surfaces are sometimes found in new or replacement work and wherever this is possible or likely, the inspector should make observation and if found, recommend the necessary changes.

Pipe Connections and Fittings. The steam and water pipes, including connections to the water column, shall be examined for leaks, and if any are found it should be determined whether they are the result of excessive strains due to expansion and contraction, or other causes. The general arrangement of the piping in regard to the provisions for expansion and drainage, as well as adequate support at the proper points, shall be carefully noted. The location of the various stop valves shall be carefully noted. The location of the various stop valves shall be observed to see that water will not be pocketed when the valves are closed and thereby establish cause for water-hammer action.

The arrangement of connections between individual boilers and the main steam header shall be especially noted to see that any change of position of the boiler, due to settling or other causes, will not produce an undue strain on the piping.

It shall be ascertained whether all pipe connections to the boiler possess the proper strength in their fastenings, whether tapped into the boiler, a fitting, or flange riveted to the boiler. The inspector shall determine whether there is proper provision for the expansion and contraction of such piping, and that there is no undue vibration tending to crystallize the parts subjected to it. This includes all steam and water pipes; and especial attention should be given to the blowoff pipes with their connections and fittings, because the expansion and contraction due to rapid changes in temperature and water-hammer action create a great strain upon the entire blowoff sytem, which is more pronounced when a number of blowoff pipes are joined in one common discharge. The freedom of the blowoff connection of each boiler shall be tested whenever possible by opening the valve for a few seconds, at which time it

can be determined whether there is excessive vibration. Blowoff pipes should be free from external dampness to prevent corrosion.

Water Column. The piping to the water column shall be carefully noted to see that there is no chance of water being pocketed in the piping forming the steam connection to the water column. The steam pipe should preferably drain toward the water column. The water pipe connection to the water column must drain toward the boiler.

Water column should be piped with cross filling whenever possible. Plugs are installed and removed to make inspections possible.

The relative position of the water column with the fire surfaces of the boiler shall be observed to determine whether the column is placed in accordance with the Code. The attachments shall be examined to determine their operating condition.

If examination is made with steam on the boiler, the water columns and gauge glasses shall be observed to see that the connections to the boilers are free as shown by the action of the water in the glass. The water columns and gauge glasses shall be blown down on each boiler to definitely determine the freedom of the connections to the boiler, as well as to see that the blowoff piping from the columns and water glasses are free. The gauge glasses shall be observed to see that they are clean and that they are properly located to permit ready observation. The freedom of the gauge cocks shall be determined by test.

Baffling--Water-Tube Boilers. In water tube boilers it should be noted as far as possible whether or not the proper baffling is in place. In many types of boilers, the absence of baffling often causes high temperatures on portions of the boiler structure which are not intended to be exposed to such temperatures, from which a dangerous condition may result. The location of combustion arches with respect to tube surfaces shall be carefully noted. These are sometimes arranged so as to cause the flame to impinge on a particular part of a boiler and produce overheating of the material and consequent danger of the rupture of the part.

Localization of Heat. Localization of heat brought about by improper or defective burner or stroker installation, or operation creating a blowpipe effect upon the boiler, shall be condemned.

Suspended Boilers--Freedom of Expansion. Where boilers are suspended, the supports and setting shall be carefully examined, especially at points where the boiler structure comes near the setting walls or floor. Often accumulation of ash and soot will bind the boiler structure at such points and produce excessive strains on the structure owing to the expansion of the parts under operating conditions.

Safety Valves. The correct pressure on the boiler is checked including Reliefing Capacity by an "Accumulation Test." If pressure is over 6%, and the maximum pressure shut off, fire immediately, open steam valve slowly, and reduce the maximum firing rate or change the safety valve to a higher reliefing capacity (lbs. per hour).

As the safety valves are the most important attachments on the boiler, they shall be inspected with the utmost caution. There should be no accumulations of rust, scale or other foreign substances located in the casings so as to interfere with the free operation of the valves. The setting and freedom of the safety valves should be tested preferably by raising the steam pressure to the blowing-off point, or if this cannot be done, the valves shall be tested by means of the try-levers to ascertain if they are free. Where the steam discharged from a safety valve is let through a pipe, the inspector shall determine at the time the valve is operating whether or not the drain opening in the discharge pipe is free and in accordance with the Code. Hand testing requires the presence of 75% pressure on the valve.

Steam Gauges. The steam gauges on all boilers shall be removed and the inspector shall test them and compare their readings with a standard test gauge. The readings of the steam gauges shall be observed and compared when making an inspection with steam on the boiler, where several boilers are in service connected to a common steam main. The location of the steam gauge shall be noted to see whether or not it is exposed to high temperature either externally, as would be the case if placed close to the smoke flue or other highly heated part of the boiler or setting, or exposed to heat internally due to lack of protection of the gauge spring with a proper siphon or trap to prevent steam from coming in contact with the spring. The inspector shall see that provisions are made for blowing out the pipe leading to the steam gauge.

If gauge is broken or not reading properly you are required to shut off the fire immediately. It is a good policy for two gauges to be installed on each boiler so that if one is not functioning properly the other one can be used.

Imperfect Repairs. Repairs are frequently improperly made, especially tube replacements. When repairs have been made, the inspector shall observe whether the workmanship is properly and safely done.

Suggestions. There are certain rules which should be followed irrespective of the type of boiler to be inspected. The inspector, whether he is the employee of a state, municipality or insurance company, should be well informed of the natural and neglectful causes of defects and deterioration of boilers. He should be conscientious and extremely careful in his observations, taking sufficient time to make the examinations thorough in every way, taking no one's statement as final as to conditions not observed by him, and in the event of inability to make thorough inspection, he should note it in his report and not accept statements of others.

Hydrostatic Testing

Knowledge of the purpose and standard practices in the application of hydrostatic tests is important to the candidate.

Study carefully the following excerpt from the New York State Boiler Code which explains the standard practice for making hydrostatic tests on a boiler pressure part.

Scope. This method of testing is applicable to materials having a definite proportional or elastic limit such as most carbon and alloy steels. It is not applicable to materials with indefinite or indeterminate proportional limits such as cast iron and most non-ferrous materials. The principle upon which the test is based assumes that the most highly stressed point in the pressure part will be subjected to a permanent set when the stress at this location reaches the proportional or elastic limit of the material. Since the stress will be directly proportional to the hydrostatic pressure, the determina-

tion of the pressure which stresses the weakest point to the proportional limit will permit a calculation of the maximum allowable working pressure that will result in a safe working stress in accordance with Code requirements for the material from which the part is made at the maximum operating temperature.

Material. The structure shall be made from material approved for its intended use by the State Boiler Code.

Workmanship. The dimensions and minimum thickness of the structure to be tested should not vary materially from those actually used. If possible, the structure to be tested should be selected at random from a quantity of such intended for use.

Preparation for Test. It is necessary to test only the weakest point of the structure, but several points should be checked to make certain that the weakest one is included. The less definte definite the location of the weakest point, the more points should be checked.

A hand test pump is satisfactory as a source of hydrostatic pressure. Either a test gauge or a reliable gauge which has been calibrated with a test gauge should be attached to the structure.

The maximum hydrostatic pressure that must be provided for will vary from 2 to 3 times the expected maximum allowable working pressure for carbon-steel structures.

The location of the weakest point of the structure may be determined by applying a thin coating of plaster of paris or similar material, and noting where the surface coating starts to break off under hydrostatic test. The coating should be allowed to dry before the test is started.

Hydrostatic Test. The first application of hydrostatic pressure need not be less than the expected maximum allowable working pressure. At least 10 separate applications of pressure, in approximately equal increments, should be made between the initial test pressure and the final pressure.

When each increment of pressure has been applied the valve between the pump and the structure should be closed and the pressure gauge watched to see that the pressure is maintained and no leakage occurs. The total deformation at the reference points should be measured and recorded and the hydrostatic pressure recorded. The pressure should then be released and each point checked for any permanent deformation which should be recorded.

Only one application of each increment of pressure is necessary.

The pressure should be increased by substantially uniform increments, and reading taken until the elastic limit of the structure has obviously been exceeded.

The pressure part shall have been subjected to a pressure greater than the designed maximum allowable working pressure prior to making the proof hydrostatic test.

Physical Characteristics of Metal.
Determine the proportional limit of the material in accordance with A.S.T.M. Specification E8-36 Standard Method of Tension Testing of Metallic Materials. It is important this this be determined from a number of specimens cut from the part tested, after the test is completed, in order to insure that the average proportional limit of the material in the part tested is used to calculate the safe working pressure. The specimens should be cut from a location where the stress during the test has not exceeded the proportional limit, so that the

specimens will be representative of the material as tested. These specimens should not be cut with a gas torch as there is danger of changing the proportional limit of the material.

Plotting Curves. A single cross-section sheet should be used for each reference point of the structure. A scale of 1 in. equals 0.01 in. deformation, and a scale of at least 1 in. equals the approximate test pressure increments has been found satisfactory. Plot two curves for each pressure point, one showing total deformation under pressure and one showing permanent deformation when the pressure is removed.

Determining Proportional Limit of Pressure Part.
Locate the proportional limit on each curve of total deformation as the point at which the total deformation ceases to be proportional directly to the hydrostatic pressure. Draw a straight line that will pass through the average of the points that lie approximately in a straight line. The proportional limit will occur at the value of hydrostatic pressure where the average curve through the points deviates from this straight line.

In pressure parts such as headers where a series of similar weak points occur, the average hydrostatic pressure corresponding to the proportional limits of the similar points may be used.

The proportional limit obtained from the curve of total deformation may be checked from the curve of permanent deformation by locating the point where the permanent deformation begins to increase regularly with further increases in pressure. Permanent deformations of a low order that occur prior to the point really corresponding to the proportional limit of the structure, resulting from the equalization of stresses and irregularities in the material, may be disregarded.

It should be made certain that the curves show the deformation of the structure and not slip or displacement of reference surfaces, gauges, or the structure.

Determining Maximum Allowable Working Pressure. (a) Having determined the proportional limit of the weakest point of the structure, the corresponding maximum allowable working pressure may be determined by the formula:

$$P = \frac{HS}{E}$$

where P = maximum allowable working pressure, lb. per sq. in.

H = hydrostatic pressure at the proportional limit of the pressure part, lb. per sq. in.

S = safe working stress permitted for the material at the maximum operating temperature as determined by Code requirements

E = average proportional limit of material, lb. per sq. in.

(b) For carbon-steel material, complying with a Code specification and with a minimum tensile strength not over 62,000 lb. per sq. in., the proportional limit may be assumed to be two-fifths the average tensile strength of the specified range. Where no range is specified, the average tensile strength may be assumed as 5000 lb. per sq. in. greater than the minimum. This will eliminate the necessity for cutting tensile specimens and determining the actual proportional limit. Under such conditions, the material in the pressure part tested should have had no appreciable cold working or other treat-

ment that would tend to raise the proportional limit above the normal value. The pressure part should preferably be normalized after forging or forming.

The safety valve pressure setting is usually the same as that found on the boiler in most localities. Furthermore, most codes indicate that "No person shall remove or tamper with any safety appliance." If you suspect a malfunction, stay with the boiler and hand operate. Observe closely and if you have any doubts concerning safety, shut down. Do not take any chances because you will be risking your license and, what is more important, the boiler, building and safety of the inhabitants of the building. Generally, you will forfeit your license if you display incompetence, negligence, or are intoxicated while on duty or for any other valid reason that establishes the operator to be unfit.

Caution. Do not install low pressure pipe or fittings on a high pressure boiler. Use only the proper material as indicated in the A.S.M.E. Pipe and Fitting Code Book. Lowering the pressure by installing a lower pressure safety valve could destroy the proper minimum relieving capacity. Perform the accumulation test to insure proper operation. Always consult the inspector before changing value.

Recommended Practices for Frequency and Method of Testing Safety Devices. The following practices are considered the minimum conducive to safe operation of equipment concerned.

Low water cutoff(s) on steam boilers should be tested not less than once a month by means of cutting off the feedwater and allowing water level to drop by evaporation, to the proper operational level of the low water fuel cutoff. The operator should be in attendance during this testing procedure. The low water fuel cutoff should also be tested at least once a week by lowering the water level in the boiler by means of the bottom blowdown on the boiler. Low water cutoff/hot water heating boilers should be tested at least bimonthly.

Safety and/or relief valves should be tested at least once a month by hand and/or preferably should be lifted by raising the pressure to the set pressure.

High limit controls should be checked for proper operation at least once a month by raising the steam pressure, or increasing the boiler water temperature above the normal operating range, to the operating level of the high limit

controls.

Flame safeguards should be tested at least monthly by "upsetting" the normal combustion of the fuel in any one of several methods which should cause the control to shut off the fuel. These safety devices should be checked for proper operation by other usual methods at least weekly.

Automatic water feed regulators should be checked for proper operation weekly. Low water cutoffs and/or feed water controls having a float chamber should be checked for proper operation and to help reduce build-up of sediment deposits by opening their drains at least weekly.

Water columns and gauge glasses should be drained at least daily.

SAFETY AT WORK

Accidents which occur at work are costly
in many ways. They cause economic hardship to the injured workers in the form
of time lost from work, medical expenses,
and human suffering. They may deprive a
a family of its breadwinner.

The majority of accidents can be
avoided. They usually fall into
one of two categories: those which
may be attributed to carelessness
either on the part of the injured
or some other worker, or those
that are caused by unsafe working
conditions. A desk drawer or file
drawer that is left open for another worker to trip over is an example of the first type. An unsafe working condition may or may
not be recognized by the employer,
but even if it is safety precautions are often not taken either
because of the expense involved or
because speed of performance may
have to be sacrificed for safety.
Some employers hope that the expense of installing a safety device or control can be avoided.
Once an accident has occurred, however, the employer soon realizes
that he has acted foolishly because
the cost of one accident is usually
much greater than many safety devices which could have been
installed to protect many workers.

Besides being expensive accidents
have an adverse affect on employee
morale. Since employee morale,
organizational efficiency and pro-

duction are directly related, one
can readily see why progressive
organizations, small and large
alike, have seen fit to safeguard
their workers while they are performing their duties.

When a worker sees accidents oc-
curring around him he is likely to
grow apprehensive. He worries that
he may fall victim to the next one.
In an effort at self-protection
he may slow down his working pace.
Sad to say, however, overly
cautious workers who perform their
duties under unsafe conditions are
not immune to accidents.

Modern personnel administration
today recognizes that workers must
perform their functions in an atmosphere permeated with safety
precautions, and that it is the
responsibility of every employer to
see that they do. Every organization should have a safety program to insure that accidents oc-
curring on its premises are kept to
a minimum. The size of the program
depends upon the type of work that
is being performed and the size of
the organization. But even in the
smallest organizations one man

should be given the responsibility for looking after the safety of the workers. It would be his function to see to it that the safety aspects of all of the jobs are looked into, and that sufficient safety guards are installed wherever a hazardous condition exists. The safety program should not be limited to locations where physical work is performed because a large number of accidents occur in offices, where one might think them highly unlikely.

It is believed by most experts in the field that there is a definite relationship between a worker's mental alertness and physical well-being and his involvement in accidents. A worker who permits his mind to wander to matters other than the work he is performing is more likely to become involved in an accident than a worker who gives his work his full concentration. A worker who is bothered by a physical ailment or who is not feeling up to par while performing his duties is also prone to accident involvement. Mentally alert workers seldom fall victim to industrial accidents. Their minds are fully occupied by the work they are doing. They are naturally careful and attentive to their surroundings.

It is impossible to implement a successful safety program without the cooperation and understanding of the workers. Signs can be posted throughout the plant attesting to the importance of the program, but if the workers are not convinced it is to their benefit to be alert and careful, they will not comply. Therefore it is imperative that every safety program, before it is implemented, have as one of its initial steps the indoctrination of the workers with the knowledge that work performance in a safe atmosphere will be of benefit to them. They must be shown how an accident is likely to cause them grief,

economic hardship and may even interfere with their career. It is is important that they understand the safety program is intended to protect their interests as well as those of the organization. A good method of getting them to accept a safety program is to permit a representative group of workers to have a say in its formation.

The responsibility for the formation of a safety program in an organization is one task that may be successfully delegated to a subordinate by the head of the agency. The individual who is chosen to conduct the safety program should be familiar with the intricacies of the various working units, and should be allotted sufficient time and help to carry out a carefully planned program. Funds should be provided for the purchase of safety devices wherever necessary, and maintenance help should be provided to eliminate dangerous situations. Loose desk lamp wires, faulty file drawers, and overloaded top drawers of standing files are examples of dangerous situations that can be easily and quickly corrected.

The actual safety program should begin with a collection of all available data concerning accidents that have occurred in the past. This data should be carefully categorized and then analyzed. Much can be learned from this information that will aid in the prevention of similar accidents in the future. Enough time should be taken for this study so that definite conclusions may be reached as to the causes of the previous accidents. The safety officer must have some "meat" with which to begin his own program. The next step should be a concentrated effort to gain the cooperation and support of all of the workers in the program. The safety officer should also consult with the various supervisors and workers throughout the organization because much can be learned from those closest to the actual work.

One of the most important aspects of a safety program is its reporting system. A report must be filled out whenever an accident occurs no matter how insignificant it may be. These reports will provide the basis for the elimination of hazardous conditions. One individual in each unit should be held responsible for completion of such reports, and he should be adequately trained to do so properly. The workers must be convinced that the sole purpose of accident reports is to bring to light hazardous conditions that may be eliminated; otherwise they may attempt to cover up small accidents which then may lead to bigger ones at some later date. The report form should be designed with simplicity in mind, but it should call for enough information so that when completed, it reflects the cause of the accident. The safety officer may find it necessary to conduct an investigation on his own in some instances, but most accident reports, if properly filled out, will form the basis for pinpointing the cause of an accident.

A safety program should be designed to bring to light hazardous conditions preferably before accidents occur, but in the event an accident does occur, the program should move swiftly to eliminate the cause of the accident so that it does not happen once more.

Attractive posters designed to alert workers to avoid dangerous situations are an integral part of a safety program. They should be placed in work locations, locker rooms, and areas where workers spend their breaks and eat their meals. The workers must constantly be reminded of the importance of safety. For example, posters depicting the proper way to lift heavy objects should be placed in strategic areas because the use of incorrect methods of lifting is the cause of many industrial accidents.

The final part of a safety program is the follow-up. It is the safety officer's job to see that his recommendations are implemented. This can prove to be quite a task. Supervisors have the job of producing maximum return from the labor and material afforded them. They may be reluctant to sacrifice speed for safety. Some supervisors may feel that "it can't happen here." The safety officer must therefore be extended sufficient authority to put his recommendations into effect. He should not resort to force, however, until absolutely necessary. The safety program will be much more beneficial if the supervisors and workers alike are convinced of its value, and are willing to cooperate with the safety officer in putting his findings to work. Therefore the safety officer's job may involve some selling, too. A safety program is bound to be successful if the supervisors, workers and safety officer all pull in the same direction.

The Accident-Prone Worker

Are individuals actually accident-prone? Are certain people more liable to involvement in accidents than others? Authorities are divided on the question. However, it is certain that some people become involved in a continuous stream of accidents whereas others avoid them almost entirely during their life span. Surely there are other factors besides lack of attentiveness and mental alertness while performing work that contribute to one's becoming involved in accidents. Some people avoid accidents by simply avoiding hazardous work situations. However, if a worker has a knack for becoming involved in accidents, how should he be treated? Should he be excused from work situations which involve any danger or should he be treated as all of the other workers are? The latest thinking on this subject places special emphasis on instructing workers who have multiple accident records. They should be taught how to work safely and to be safety conscious. An organization could not function properly if assignment of personnel were determined solely by the individual's accident record. A worker must perform at the job that will best suit the organization.

If a worker continues to become involved in accidents after he has been sufficiently trained and indoctrinated in safe work procedures, consideration should be given to discharging him. Most workers can avoid accidents if they put their minds to it. Special assignments for workers should only be made at the convenience of the organization.

Safety Awards

Although abstinence from accidents is reward in itself for a working unit, most safety programs will also issue formal awards to accident-free groups in an organization. These awards are usually made in certificate form so that they can be displayed within the working area of a unit.

These awards are more accurately certificates of recognitiion. They serve to indicate that a working unit has recognized the value of safe working procedures to themselves and the organization alike. These units have seen fit to comply with the rules of safety in order to promote organizational efficiency, and to rid the workers of an evil to which they do not have to fall victim.

These certificates of recognition tend to take an added importance in time. For a working unit not to po possess a safety certificate may symbolize more than a poor safety record. It will indicate a lack of desire on the unit's part to do what has been deemed best for both themselves and the organization. The certificates will then become something that every unit will strive to attain to show that they want to "join the team." The competition for acquisition of safety certificates will indirectly achieve the result that the entire safety program was intended for: an accident-free organization with increased efficiency.

PRACTICE QUESTIONS

1. If a new employee becomes involved in an accident he should be

   (A) given additional training in safe working methods
   (B) discharged immediately
   (C) held strictly responsible for the accident
   (D) no longer assigned to tasks where there is any danger involved.

2. One of the workers under your supervision asks to be excused from performing a task on a particular day because it involves the use of a 12-foot ladder. He has performed this job before on numerous occasions, but he says he is not feeling just right today. You should

   (A) insist that he perform the function
   (B) tell him to find another worker who is willing to change places with him
   (C) assign the task to another employee this time
   (D) report him to your superior.

3. If a worker continues to perform a job in an unsafe manner after he has been trained by his supervisor to perform it in the proper manner, he should be

   (A) permitted to perform the job in his own way because it is easier for him to do so
   (B) disciplined
   (C) praised because he has not yet become involved in an accident
   (D) once again retrained by his supervisor.

4. In an accident report the <u>least</u> useful of the following information would be the

   (A) name of the injured worker

   (B) time and date accident occurred
   (C) length of time injured worker has been employed by the agency
   (D) extent of the injury.

5. A supervisor who observes one of his subordinates performing his duties in an unsafe manner should

   (A) wait until the next staff meeting to correct him so that the whole staff will benefit
   (B) say nothing at the time, but wait until he observes him doing something else wrong to bring it up
   (C) have another subordinate correct him at the first opportunity
   (D) correct him at once.

6. A report of an accident should be made as soon after the occurrence as practical chiefly because

   (A) the witnesses may change their stories before the report is written
   (B) the reporter should strive to rid himself of the chore so he can devote himself to more important work
   (C) the details of the accident will still be fresh in the reporter's mind
   (D) reports should be submitted promptly even if they do not contain all of the facts.

7. If most accidents in a unit whose workers are performing routine work of a somewhat hazardous nature occur in the last hour of the workday, they are most likely due to

   (A) carelessness on the part of the workers
   (B) improper training of the workers
   (C) lack of interest in safety on the part of the workers

(D) mental and physical
fatigue on the part of the
workers.

8. If one of your subordinates
approaches you, a supervisor,
with the information that one
of the functions he is performing may cause an accident,
you should tell him

(A) you are the supervisor and
it is your function to
determine work procedures
(B) to put his information in
writing
(C) to refer the information
to the safety officer
(D) that you will investigate
the matter at once.

9. If, after completing a training course in safe work
procedures, the trainees still
continue to use the same procedures they used before
attending the course the most
probable reason is that

(A) they were not convinced
of the value of the pro-
cedures taught to them in
the course
(B) they have not absorbed
the training material
(C) the training material was
not pertinent
(D) they are willing to engage
in dangerous practices in
order to get the work done
quickly.

10. If you as a supervisor observe
a worker not under your jurisdiction performing his duties
in an unsafe manner you should

(A) immediately instruct him
in the proper method of
work performance
(B) ignore him because the
worker is not under your
jurisdiction
(C) seek out his supervisor
and inform him of his
subordinate's action
(D) report the worker to the
safety officer.

11. If you observe what seems to be
a serious accident to a worker
in your unit, your first
course of action should be to

(A) seek out the supervisor
(B) fill out an accident report
(C) notify the injured worker's
family
(D) render any possible first
aid to the victim.

12. It is a poor practice to fill
the top drawer of a file cabinet entirely before using the
drawers beneath it chiefly
because

(A) some file clerks may be
shorter than others
(B) the cabinet will topple
forward when the top
drawer is opened entirely
(C) the file is not being used
to its utmost capacity
(D) material should be distributed equally throughout
the file.

13. In order to avoid injuries it
is proper to lift heavy objects

(A) with the knees bent and
the back held stiff
(B) with both the knees and
the back held stiff
(C) using the arms only
(D) in the manner best suited
to the individual doing the
lifting.

14. The least likely reason for including a safety program in an
organizational setup is

(A) the type of work being
performed
(B) the number of workers
employed
(C) the number of levels in
the organization
(D) the time lost from work due
to accidents.

15. A safety officer can best secure
the cooperation in a program
which he is conducting if he

(A) exercises the authority extended to him by the head of the agency

(B) waits until a serious accident occurs to point out the necessity of a safety program

(C) strives to become "one of the boys"

(D) is able to convince the workers that the safety program is to their best interests.

16. If a worker who is to perform a job in a hurry notices that a piece of equipment he is to use is defective, he should

(A) hold off completing the job until the equipment is either replaced or repaired

(B) go ahead with the job because it must be completed in a hurry

(C) consult with another subordinate

(D) be cautious while using the defective equipment.

17. A worker notices a piece of equipment which he believes can aid him in performing his work. The equipment is somewhat complicated, and he has never received any instruction as to its operation, although other employees have. He should

(A) use the equipment because he will be able to do his work faster

(B) put off doing his work until he has received adequate instruction in the use of the equipment

(C) ask another worker to teach him to use the equipment

(D) not use the equipment until he has received adequate instruction in its operation.

18. If after completing an accident report you discover an additional piece of information which may influence the conclusions drawn from the report you should

(A) ignore the information because the report has already been completed

(B) rewrite the report including the additional information you have discovered

(C) submit the original report and wait to see if the additional information is pertinent

(D) ask the injured worker if he thinks it should be included in the accident report.

19. If you find that your subordinates are using improper tools while performing their functions, and if, after confronting them with this information, they tell you that requisitioning the proper tools from the tool shack is too time consuming, you should

(A) insist they use the proper tools

(B) go along with them because they have performed well in the past

(C) inform them that they will be held strictly responsible for any accidents that occur from their use of the improper tools

(D) report them to their supervisor.

20. When a veteran worker with a clean safety record suddenly becomes involved in a series of work accidents it would be wise for the supervisor to first

(A) retrain the employee in safety procedures

(B) discipline the employee

(C) try to discover the cause of the accidents

(D) transfer the employee to another unit.

Explanatory Answers

1. (A) The new worker has demonstrated that he may not yet be familiar with all of the safety aspects of his duties. This is the time to give him additional training and observe how he responds.

2. (C) There is nothing in the question to make one believe this worker is a malingerer. It is reasonable for the supervisor to grant his request.

3. (B) This employee is disregarding specific instructions given to him by his supervisor. He should be disciplined immediately before he becomes involved in an accident. Once he has, it will be too late.

4. (C) The primary purpose of an accident report is to isolate the cause of the accident so that proper remedial action can be taken so that it does not occur again. Choices (A), (B), and (D) would aid in determining the cause of the accident. The length of time the injured worker has been employed by the agency would have very little bearing.

5. (D) There is no point in delaying the correction of a subordinate who is performing his duties in an unsafe manner. He may fall victim to an accident in the meantime. The worker should be corrected immediately.

6. (C) Accident reports should be prepared promptly because the facts will still be fresh in the mind of the reporter, and he will be less likely to omit important items.

7. (D) When accidents occur mostly at the end of a workday they can most likely be attributed to mental and physical fatigue on the part of the workers, especially where routine work is being performed. Perhaps more work breaks would alleviate the situation.

8. (D) Often those closest to the actual work will see things in a job the supervisor will overlook. The supervisor should always keep an open mind to suggestions from his subordinates.

9. (A) If trainees do not use the material taught them in a training session, they are probably not convinced of the value of the training material.

10. (C) It would be proper for you to talk to his supervisor first. Both the worker and the supervisor may resent any direct contact you may have with the worker because he is actually not responsible to you.

11. (D) In this situation all formalities should be abandoned. Your first action would be to render any immediate help to the victim that you can.

12. (B) If a file cabinet is loaded from the top down, it will topple forward when the top drawer is opened all the way. This is a common cause of office accidents.

13. (A) When this method is used, the stress is being equally distributed between the back and the legs.

14. (C) The number of levels in an organization would not have any bearing at all on the necessity of a safety program.

The other three choices do have a bearing.

15. (D) The safety officer will be most able to secure the cooperation of the workers in an organization if he can show them how it will benefit them, and how they stand to suffer if they are involved in an accident.

16. (A) The use of defective equipment is the cause of many an accident. Unfortunately, caution will not always prevent accidents if dangerous equipment is utilized.

17. (D) It is wrong for a worker to use any piece of equipment on which he has not been adequately trained. If the supervisor had wanted him to use the equipment in question, he would have seen that he was trained in its use.

18. (B) An incomplete report may become a harmful instrument.

19. (A) The supervisor should insist that his subordinates use proper tools no matter what inconveniences they may encounter in securing them. The use of improper tools has caused many accidents.

20. (C) This veteran employee has demonstrated through long service without accidents that he knows how to work safely. Obviously something is bothering him; it is the supervisor's job to discover what it is.